____The____
Subcritical Brain

A Synergy of Segregated Neural Circuits in
Memory, Cognition and Sensorimotor Control

The
Subcritical Brain

A Synergy of Segregated Neural Circuits in Memory, Cognition and Sensorimotor Control

Yoram Baram

Technion – Israel Institute of Technology, Israel

World Scientific

NEW JERSEY · LONDON · SINGAPORE · BEIJING · SHANGHAI · HONG KONG · TAIPEI · CHENNAI · TOKYO

Published by

World Scientific Publishing Co. Pte. Ltd.

5 Toh Tuck Link, Singapore 596224

USA office: 27 Warren Street, Suite 401-402, Hackensack, NJ 07601

UK office: 57 Shelton Street, Covent Garden, London WC2H 9HE

Library of Congress Cataloging-in-Publication Data

Names: Baram, Yoram, author.

Title: The subcritical brain : a synergy of segregated neural circuits in memory, cognition and
sensorimotor control / Yoram Baram, Technion - Israel Institute of Technology, Israel.

Description: New Jersey : World Scientific, [2021] | Includes bibliographical references and index.

Identifiers: LCCN 2020055843 | ISBN 9789811233098 (hardcover) |
ISBN 9789811233104 (ebook for institutions) | ISBN 9789811233111 (ebook for individuals)

Subjects: LCSH: Neural networks (Computer science) | Neural computers--Circuits.

Classification: LCC QA76.87 .B36 2021 | DDC 612.8--dc23

LC record available at https://lccn.loc.gov/2020055843

British Library Cataloguing-in-Publication Data

A catalogue record for this book is available from the British Library.

For any available supplementary material, please visit
https://www.worldscientific.com/worldscibooks/10.1142/12182#t=suppl

Typeset by Stallion Press
Email: enquiries@stallionpress.com

I am deeply indebted to Ophira, Tami, Orli, Amir, Ur, Maya, Lotem, Sara, Ya'ar and Arava for being there

The Subcritical Brain weaves together theoretical ideas from random graphs, nonlinear dynamics, prime numbers and quantum computation, to address one of the most important questions in science — how brains compute. A must read for anyone interested in theoretical studies of cortical microcircuits.

—Prof Dario Ringach, Professor of Neurobiology & Psychology, David Geffen School of Medicine, University of California, Los Angeles

The field of Neural Networks has grown in the 1980s, and became a very active interdisciplinary arena leading to new insights into two major research topics: brain research and computational modeling. The latter has developed into Machine Learning, branching out recently into Deep Learning.

This book represents very well this dual character of the field. In some chapters, the author draws on lessons from Mathematics and Engineering, to explain or paraphrase brain processing. In others he applies lessons from Biology to the construction of novel engineering and computational devices and techniques.

Any reader who is fond of Applied Mathematics will be delighted to witness the author's skillful use of abstract tools in investigating brain mechanisms. His discussion follows many of his original contributions to the field. In the last chapters he summarizes his work with leading neurologists in addressing issues of movement disorders in neurologically impaired patients.

Baram's main thesis is that weakly connected small neural circuits offer higher information processing capacities than large neural networks. Paraphrasing this message we may emphasize the importance of diversity in both the brain's system of systems and the education of its researchers. The book is a fine example of the latter, widening the scope of the student and the researcher who wishes to augment his/her education on a multitude of pertinent topics.

—Prof David Horn, Professor of Physics, Incumbent of the Edouard and Francoise Jaupart Chair of Theoretical Physics of Particles and Fields, Tel Aviv University

Professor Yoram Baram has dared set one foot in the world of advanced mathematics and the other in the world of medical practice. In *The Subcritical Brain*, he creates a bridge between these two worlds that a reader can walk with both confidence and amazement. Neural Networks may have never been approached with such insight and lyricism.

—Prof Alberto Espay, Professor of Neurology, University of Cincinnati

Few areas in science have attracted the huge amount of research as the human brain.

The Subcritical Brain provides a translational mathematical framework for understanding the neural systems underlying the human mind and behavior and consequently the clinical sciences of neurology and psychiatry.

The content covers the historical development of the science of modeling biological networks mathematically, from its birth through the most updated research methods and understandings.

All chapters begin with a short mention of the behavioral-neurophysiological process, continue with the presentation of the prevailing mathematical and theoretical positions that reflect the diversity of approaches to the addressed subject and culminate in a detailed integrative analysis of the author's insights and research results.

All chapters are of high scientific and literal quality. The book is recommended to all those interested in translational science aiming to model the abilities of the human mind which is suitable to adapt to our ever-changing environment.

—Prof Judith Aharon-Peretz, Professor of Neurology, Medical School, Technion – Israel Institute of Technology

Prologue

My first acquaintance with the topic of neural networks in the late 1980s has taken me to the question of cortical information capacity. Networks of many neurons, communicating in a binary (on/off) fashion, have appeared to represent a biological reality and seemed reasonably easy to address mathematically. Replacing "many" by "infinitely many" seemed an acceptable mathematical convenience. Having read a few papers by others, and having published a few of my own, I was left with a sense of discomfort. The notion of an infinite-size neural network, marginally surviving in a contrived reality, became an annoying mental burden. It certainly did not match the finite size of the brain, any brain. Turning to issues of sensorimotor control in the neurologically impaired has made for a decade of satisfying self-fulfillment.

My return to more realistically founded cortical mathematics was sparked by a realization that it is the dynamics of change, not stationary binary state, which define cortically governed behavior. This realization was highly enhanced by the relationship between seemingly different mathematical notions: global attractors, which define non-invertible firing-rate dynamics; random graphs, which define neural circuit connectivity; and prime numbers, which define the dimension and category of cortical operation. An intriguing aspect was offered by the application of quantum computation.

Somewhat surprisingly, while quantum mechanics often yields fundamentally different results than classical mechanics, quantum computation — shown to be embedded in certain cortical operations — has yielded the same key conclusion as classical computation: weakly connected small neural circuits facilitate higher information storage and processing capacities than large circuits, which, by random graph considerations, tend to be highly connected. The conceptual, linguistic and functional burdens of infinite-size neural networks were swept away. The complementary exposition to the cortical origins of such wondrous natural phenomena as musical perception, bird descending trajectory, and, in particular, neurological patient struggle, resolve and achievement, has made this journey endlessly gratifying.

I have greatly benefitted from the wisdom I found in the scientific works of others, and am indebted to those individuals who have helped me realize that wisdom in my own theoretical and practical endeavors: Ralph Abraham, Judith Aharon-Peretz, Amir Baram, Ran El-Yaniv, Yuval Filmus, Alberto Espay, Ruben Lenger, Ariel Miller, Nathan Peterfreund, Dario Ringach, Lior Ron, Ze'ev Roth, Virginia de Sa, Dror Sal'ee, Tamer Salman, Yahalomit Simionovici, Priya Velu and Mark Zlochin. Many other collaborators are noted as co-authors of articles referenced throughout this book.

Yoram Baram
Haifa, 2021

Contents

Part IV. Sensorimotor Control 213

Chapter 15. Circuit Polarity in Sensorimotor Control 215

Chapter 16. Autonomous Gait Entrainment in the Neurologically Impaired 227

Chapter 17. Circuit Polarity and Singularity Segregation in Cortical Recordings 241

Chapter 1

Introduction

1.1 General

Traditionally, the experimental nature of biological research has required a high degree of phenomenological specificity. Consequently, seemingly separate issues such as cortical development, cortical functions, inter-neuron connectivity, neuronal firing dynamics, learning and memory have been studied in seemingly complete mutual isolation. Yet, recent years have seen a trend towards integrative neuroscience, driven by a realization that, on the one hand, practically every cortical function combines a variety of molecular, biological and physical processes, and, on the other, different cortical functions are often executed by similar mechanisms. It is also becoming increasingly clear that a complete understanding of such integration would involve a new look at the cortical arena, incorporating advanced theoretical reasoning and new methods of analysis. The relatively recent emergence of numerous publications presenting advanced mathematical analysis of neurobiological structures and processes is strong evidence of this effect. Recent interdisciplinary interest in general common notions such as information, linguistics, learning, memory, computation, cognition, dynamics and control is further evidence of the same effect. It is the purpose of this book to address these issues in an analytically rigorous manner, combining relevant experimental findings and advanced theoretical concepts. While the general methodology and many of its consequential outcomes are original, they are supported by several recent publications by the author along with a wide range of references to publications by others, old and recent, related to the issues under consideration.

1.2   Historical account

1.2.1   *The nature of cortical information*

The nature of cortically embedded information has been intensely studied for over a century. Yet, it remains one of the most highly evasive mysteries of neuroscience. Following Cajal's introduction of the neuron in the late 19th century (e.g., Cajal, 1890), early investigations of cortically embedded information have concerned the nature of neuronal firing (Lapicque, 1907). Elaborate chemical, physical and mathematical analysis has led to the celebrated conductance-based model of action potential (Hodgkin and Huxley, 1952). The dynamic nature of empirically observed trains of neuronal firing impulses (Adrian and Zotterman, 1926) has remained, however, more enigmatic. The very need for variability in the neuronal firing dynamics has been questioned (Stein *et al.*, 2005; Faisal *et al.*, 2008). While empirically observed firing sequences have been described as tonic (Murthy and Fetz, 1996; Bennett *et al.*, 2000), oscillatory (Murthy & Fetz, 1996; Elson *et al.*, 2002; Cymbalyuk and Shilnikov 2005; Wang, 2010), periodic (So *et al.*, 1998), quasi-periodic (Lankheet *et al.*, 2012), multiplexed (Panzeri *et al.*, 2009), or silent (Melnick, 1994; Epsztein *et al.*, 2011), some have been characterized as random (Gerstein and Mandelbrot, 1964), or as chaotic, representing functional bursting (Hayashi and Ishizuka, 1992), or deficient states of neural information processing (Fell *et al.*, 1993). Individual neurons of the same type are often capable of producing different firing modes, switching from one to another in a seemingly unpredictable manner (Hyland *et al.*, 2002). However, the mathematical rules underlying such behavior, the purpose it might serve, or the harm it might cause, have not been well understood. The transition from one dynamic mode to another has been called global bifurcation when caused by the landscape of the underlying map subject to fixed parameter values, and local bifurcation when caused by a change in parameter values (Blanchard *et al.*, 2006). Bifurcations with chaos (Ren *et al.*, 1997) and bifurcations without chaos (Li *et al.*, 2004) have been reported. Biologically-based, analytically derived bifurcation models have been shown to produce spiking and bursting sequences (Izhikevich, 2000, Kuznetsov *et al.*, 2006). Periodic bifurcations may in themselves represent a dynamic mode, as in the case of periodic bursting (Elson *et al.* 2002) instigated by postinhibitory rebound (postinhibitory facilitation, Perkel and Mulloney 1974). The choice between the spiking paradigm, putting the spotlight on the changing time intervals between neuronal spiking, and the firing-rate paradigm, seeking the information in the dynamics of interval-averaged number of spikes has been highly debated (Gerstner and Kistler, 2002). The firing-rate paradigm, while lacking in detail with respect to the spiking paradigm, has been found to offer a certain mathematical convenience, and, at the same time, reliably reproduce empirically observed firing sequences (Wilson and Cowan, 1972; Gerstner, 1995; Dayan and Abbott, 2001; Jolivet *et al.*, 2004). Following a mathematical analysis of discrete iteration maps for neural network firing-rate, a global attractor code, relating the modes of firing-rate dynamics to internal neuron properties, has been derived (Baram, 2012, 2013a).

1.2.2 *Developmental aspects*

Neurophysiological and molecular studies have distinguished between pre-critical excitability, instrumental in initial circuit formation (Hsia *et al.*, 1998; Hensch *et al.*, 1998; Hensch, 2005; Ashby and Isaac, 2011) and persistent plasticity, evidenced during critical development of ocular dominance in early life (Hensch, 2005; Miyata *et al.*, 2012). Empirical characterizations of developmental stages have employed sensory responses. High sensitivity to sensory experience is crucial for early neural circuit formation (Hooks and Chen, 2007). Excitatory plasticity in pre-critical period immediately following birth has been found to govern activity independent synapse and dendrite generation and arborization, and axon growth, branching and targeting (Hsia *et al.*, 1998; Ashby and Isaac, 2011; Tessier and Broadie,

2009; Gibson and Ma, 2011; Weiner *et al.*, 2013). At the same time, the Hebbian paradigm supports activity-dependent inter-neural connectivity (Caporale and Yang, 2008; Buzsáki, 2010; Doll and Broadie, 2014). Molecular mechanisms for cortical plasticity control have been suggested, with both excitatory and inhibitory synapses playing critical roles (Constantine-Paton *et al.*, 1990; Huang *et al.*, 1999; Fagiolini and Hensch, 2000; Goold and Nicoll, 2010; Phillips *et al.*, 2011; Miyata *et al.*, 2012; Wang *et al.*, 2012). It is largely believed that initial excitability enhances sensitivity to inhibitory effects, holding the key to plasticity modification (Huang *et al.*, 1999; Fagiolini and Hensch, 2000; Feller and Scanziani, 2005; Goold and Nicoll, 2010; Wang *et al.*, 2012). Beyond activity-independent neural circuit formation in early development (Hsia *et al.*, 1998; Gibson and Ma, 2011; Weiner *et al.*, 2013), activity-dependent circuit formation (Katz and Shatz, 1996; Gage, 2002) and later modulation (Tessier and Broadie, 2009), neural circuit modification by activation and silencing of neuronal membrane and individual synapses has been observed in early development (Melnick, 1994; Atwood and Wojtowicz, 1999; Liao *et al.*, 1999; Losi *et al.*, 2002; Kerchner and Nicoll, 2008), maturation (Ashby and Isaac, 2011) and later life (McGahon *et al.*, 1999). The distinction between developmental stages has not been matched by the mathematical representation of cortical plasticity on the one hand, and of neural firing, on the other. Yet, a generalized developmental approach to biologically realistic mathematical modeling and prediction of neural connectivity has been proposed, relating neural network growth to activity and function (Borisyuk *et al.*, 2011, 2014).

1.2.3   *Cortical connectivity*

Cortical circuit connectivity has attracted increasingly growing interest for the past three decades. Sensory inputs have been shown to evoke ongoing shunting in visual cortex circuits (Borg-Graham *et al.*, 1998). Activity-independent neural circuit formation in early development (Hsia *et al.*, 1998; Gibson and Ma, 2011; Weiner *et al.*, 2013) has been found to be followed by activity-dependent circuit formation (Katz and Shatz, 1996; Gage, 2002) and later modulation

(Tessier and Broadie, 2009). Neural circuit modification by activation and silencing of neuronal membrane and individual synapses has been observed in early development (Melnick, 1994; Atwood and Wojtowicz, 1999; Liao *et al.*, 1999; Losi *et al.*, 2002; Kerchner and Nicoll, 2008), maturation (Ashby and Isaac, 2011) and later life (McGahon *et al.*, 1999). Cortical segregation into small groups of neurons has been related by simulation to radius of inhibition and found to have an effect on spiking dynamics, leaving a deeper understanding of the observed activity for future research (Stratton and Wiles, 2015). Synchronous and asynchronous reverberation have been synthetically embedded in local cortical circuits and individual neurons (Vardi *et al.*, 2012).

1.2.4 *Somatic and synaptic polarization*

The state of neuronal activity, contrasted by neuronal silence, has been found to depend on the somatic membrane potential being above or below a certain threshold value (about -60 mV, Melnick, 1994). The state of synaptic transmissivity, contrasted by synaptic silence, has been found to depend on the value of pre-synaptic membrane potential, controlled by external stimulation, and molecular properties with respect to a certain threshold value (also about -60 mV, Atwood and Wojtowicz, 1999). A detailed biophysical model relates long-term synaptic potentiation and long-term synaptic depression, which are viewed in a binary ("bidirectional") context, to the variable properties and relative numbers of AMPA (α-amino-3-hydroxy-5-methyl-4-isoxazolepropionic acid) and NMDA (N-methyl-D-aspartate) receptors and their external stimulation (Castellani *et al.*, 2001). As the definitions of membrane and synapse polarities have been derived directly from experimental neurobiological findings independently of any particular firing or plasticity models, we are able to address neuronal polarity gates, neural circuit polarity codes and the corresponding issue of circuit segregation in a discrete mathematical framework independent of such models. While the memory mechanization by circuit polarization proposed in the present work following Baram (2018) does not rule out a certain role for the death and regrowth of neurons and synapses in the implementation of memory, it appears to be considerably more economical, controllable and agile than the latter.

1.2.5   *Synaptic and somatic elimination*

Permanent synaptic and axonal elimination have been observed in humans (Huttenlocher, 1979; Huttenlocher *et al.*, 1982; Huttenlocher and Courten, 1987) and in animals (Eckenhoff and Rakic, 1991; Bourgeois, 1993; Bourgeois and Rakic, 1993; Rakic *et al.*, 1994; Innocenti, 1995). Perceived as the removal or "pruning" of redundant or weak synapses for the improvement of neural circuit performance, such structural circuit modification has been suggested as an ongoing procedure for grey matter maintenance and upkeeping (Balice-Gordon and Lichtman, 1994). While early studies have associated synapse elimination with early development (Balice-Gordon and Lichtman, 1994; Culican *et al.*, 1998) and childhood (Chechik *et al.*, 1998), others have extended it to puberty (Iglesias *et al.*, 2005) and, depending on brain regions, to age 12 for frontal and parietal lobes, to age 16 for the temporal lobe, and to age 20 for the occipital lobe (Giedd *et al.*, 1999). Yet, Alzheimer's disease (Horn *et al.* 1996), grey matter (Mechelli *et al.*, 2004) and cognition (Craik and Bialystok, 2006) studies, and persistent evidence of molecular processes involved in synaptic elimination throughout life (Lee *et al.*, 2016) have suggested its continued relevance. Connectivity changes due to synapse elimination (Dennis and Yip, 1978; Huttenlocher, 1979) have been suggested as means for long-term memory (Balice-Gordon *et al.*, 1993), supported by studies of structure (Balice-Gordon and Lichtman, 1994; Chklovsii *et al.*, 2004; Knoblauch and Sommer, 2016), information capacity (Knoblauch and Sommer, 2016) and cortical segregation (Baram, 2017b).

1.2.6   *Cortical plasticity*

Cortical plasticity, underlying the brain's ability to change, has been largely viewed in two separate contexts: early development, governed largely by genetics; and later adaptation, governed largely by experience (Feller and Scanziani, 2005). Activity-dependent changes in synaptic plasticity ("the plasticity of synaptic plasticity") have been termed "metaplasticity" (Abraham and Bear, 1996; Abraham, 2008; Ming-Chia *et al.*, 2010). Age-related cortical degradation is yet another aspect of plasticity, or its demise. Cortical plasticity has physical, chemical and biological expressions. Fundamental changes in cortical modes of firing are perhaps the most highly visible expressions of cortical plasticity. Synaptic plasticity, believed to affect and be affected by neuronal firing underlying learning and memory in the nervous system (Hebb 1949; Bienenstock *et al.*, 1982; Dudai, 1989; Cooper *et al.*, 2004) has been assumed to evolve on a slower time scale, often separated from firing dynamics for analytic convenience. While almost all theoretical and experimental studies make the implicit assumption that synaptic efficacy is both necessary and sufficient to account for learning and memory, it has been suggested that learning and memory in neural networks result from an ongoing interplay between changes in synaptic efficacy and intrinsic membrane properties (Marder *et al.*, 1996). A filtering property of the neuron, band-passing its own feedback and inputs from interacting neurons with the same behavior, is implied by the Hebbian paradigm (Hebb, 1949) and supported by the eigen-frequency preference paradigm of spiking neurons (Izhikevich, 2001). Experimental imaging relating circuit formation to coordinated neural firing activity (Kenet *et al.*, 2003; Karlsson and Frank, 2009; Komiyama *et al.*, 2010; Garner and Mayford, 2012) further supports the mutual filtering paradigm among interacting neurons. Synchrony has been observed in both small circuits (Komiyama *et al.*, 2010; Marder and Bucher, 2001) and larger structures, even whole brain tissues (e.g., Lopes da Silva, 1991).

A detailed biophysical model of long-term synaptic potentiation and long-term synaptic depression has been presented (Castellani *et al.*, 2001), supporting the BCM plasticity theory (Bienenstock *et al.*, 1982). The combined effects of firing-rate and plasticity time constants on firing-rate dynamics corresponding to different developmental stages have been analyzed, laying the ground for a firing-rate dynamics-based theory of metaplasticity, ranging across neuronal properties on the one hand, and cortical development on the other (Baram, 2017a).

1.2.7 *Learning and memory*

Learning and memory are unquestionably among the most fundamental concepts associated with advanced living objects. While human awareness of learning and memory is conceivably as old as humanity itself, their mathematically formal conceptualization has seemed unattainable until only a few decades ago. The following writing by a great 17th century mathematician, Descartes, has been cited (e.g., Dudai, 1989) as the first informal conceptualization of memory on record:

> *Thus, when the soul wants to remember something* $\cdots$ *volition makes the gland lean first to one side and then to another, thus driving the spirits towards different regions of the brain until they come upon the one containing traces of the object we want to remember. These traces consist simply of the fact that the pores of the brain through which the spirits previously made their way, owing to the presence of this object, have thereby become more apt than the others to be opened in the same way when the spirits again flow towards them. And so the spirits enter into these pores more easily when they come upon them, thereby producing in the gland that special movement which represents the same object to the soul, and makes it recognize the object as the one it wishes to remember. (Descartes, 1649).*[1]

Culturally, the distinction between learning and memory has been based on behavioral attributes such as the level of effort or the length of time involved. However, by measure of outcome,

---

[1]It is quite coincidental that Descartes, by his widely known results such as the rule of signs (Fine and Rosenberg, 1997), has made it possible for this author to mathematically prove the non-invertibility of the neural firing-rate process (see Chapter 7; Baram, 2012). This has, in turn, motivated the attractor approach taken in subsequent works and in this book.

there does not seem to be any difference between learning and memory at all. The first formal (yet, not mathematical) conception of biological learning and memory was presented by Hebb (1949), whose poetically phrased concept "neurons that fire together wire together" has inspired many mathematical variants of this paradigm. An early biologically-inspired learning mechanism incorporating Hebb's paradigm has been termed *perceptron* (Rosenblatt, 1958). Mathematically founded on solution-finding convergence, the perceptron has been found to be limited to linearly separable information (Minsky and Papert, 1969). Yet, it has laid the ground for a powerful theory of machine learning (Vapnik, 1995) and its statistical ramifications (e.g., Roth and Baram, 1996; Zlochin and Baram, 2001). Progress towards a biologically applicable theory on learning and memory has been more modest. Attracting considerable interest in mathematical dynamics and information theoretic circles, it has been linked to the convergence and stability of neural network activity models with respect to parametrically stored states. However, the analysis of such models in discrete (McCulloch and Pitts, 1943; Amari, 1972; Hopfield, 1982) and continuous (Cohen and Grossberg, 1983; Peterfreund & Baram, 1998a,b) space and time has presented a sharp tradeoff between low storage capacity, sublinear in the number of neurons (McEliece *et al.*, 1987; Kuh and Dickinson, 1989; Dembo, 1989), and a multitude of spurious outcomes (Bruck and Roychowdhury, 1990). Sparse distribution of active neurons has been shown to raise the sublinear capacity bound (Baram and Sal'ee, 1992) while an exponential increase in the number of neurons (Kanerva, 1988) has been shown to result in exponential growth in storage capacity with respect to the dimension of the data stored (Chou, 1989). The mechanization of the associative memory concept presented in these works seems to have fallen short of a widely acceptable biological support. The biologically motivated mathematical approach taken in this book makes a very slight distinction between learning and memory, which, adhering to the notion of firing-rate dynamics-based metaplasticity (Baram 2017a), associates the former with the formation of a certain dynamic mode of neural firing-rate and the latter with the outcome of such formation.

1.2.8 *Cortical function*

While there seems to be a clear relationship between certain firing modes and neural functions (e.g., oscillation (Sharp *et al.*, 1996) appears directly related to heartbeat, walking and chewing), the utility of others is not as commonly recognized or understood. Chaotic neural firing has been conjectured to represent functional pace-making by rhythmic bursting (Hayashi and Ishizuka, 1992) and deficient states of neural information processing (Fell *et al.*, 1993). Temporal multiplexing (i.e. transmitting and receiving independent signals over a common signal path) of different firing signals, analytically modelled (Izhikevich, 2001) and observed in sensory cortices (Fairhall *et al.*, 2001; Wark *et al.*, 2009; Lundstrom and Fairhall, 2006), enhances the coding and information transmission capacity (Bullock, 1997; Lisman and Grace, 2005; Kayser *et al.*, 2009). Although temporal precision of the multiplexing code can be achieved by narrow windowing, the need for such precision in neural coding has been questioned (Panzeri *et al.*, 2009). While a chaotic attractor drives temporal mixing of firing-rates over the entire state-space, a largely cyclic attractor can perform multiplexing of two oscillatory signals. Depending on the function of the receiving neuron, demultiplexing can be done, in principle, by band-pass filtering. Neuronal low-pass (Pettersen and Einevoll, 2008) and high-pass (Poon *et al.*, 2000) filtering have been reported. Yet, the raw multiplexed, and even chaotic, signal can be useful in sensory systems. Multiplexed Red, Green and Blue (RGB) color coding is a known example of creating mixtures of the primary colors, found in both biological and technological vision systems (Hunt, 2004). A silent attractor, representing the state of a silent neuron, has been found to play a major role in cortical representation of place, largely termed *place neurons* (O'Keefe and Dostrovsky, 1971; O'Keefe and Nadel, 1978; Calvin, 1996; Hafting *et al.*, 2005; Ashby and Isaac, 2011; Epsztein *et al.*, 2011). More generally, sensorimotor control, fundamental to life and survival of both the individual and the species, is perhaps the most essential of all cortical functions. A synergy between neurobiological research and clinical studies has produced major advances in the understanding of normal sensorimotor control as well as medical procedures for the neurologically impaired (e.g., Baram, 2013b).

1.2.9 *Graphs and categories*

A paper written by Leonhard Euler on the Seven Bridges of Königsberg and published in 1736 is regarded as the first paper in the history of graph theory (Biggs *et al.*, 1986). Euler's formula relating the number of edges, vertices and faces of a convex polyhedron was studied and generalized by Cauchy (1813) and L'Huilier (1812–1813) and represents the beginning of the branch of mathematics known as topology.

The earliest use of a random graph model compared the fraction of reciprocated links in network data with a random model (Moreno & Jennings, 1938). Another use employed a model of directed graphs with fixed out-degree and randomly chosen attachments to other vertices (Solomonoff and Rapoport, 1951). The most widely recognized models of random graphs (Erdos and Rényi, 1959, 1960; Gilbert, 1959) were motivated by the emergence of computer communication systems. Although neural circuits and networks can be graphically described, powerful graph theoretic results remain essentially in the abstract mathematical domain, escaping the attention of neuroscientists for many years.

The concepts of category and categorization — the grouping of objects — have been at the center of philosophy, linguistics and science at large for several millennia. Graphs are mathematical structures used to model pairwise relations between objects (Trudeau, 1993). Category theory (Awodey, 2010) has engaged the concept of directed graphs in formalizing mathematical structures of objects sharing certain attributes. The "arrows" (or "morphism") of such graphs are often said to represent a process connecting two objects. An "arrow" connecting an object to itself is called an "identity morphism". An early application of category theory outside pure mathematics appears in the "metabolism-repair" model of autonomous living organisms (Rosen, 1958).

1.2.10 *Prime numbers*

A prime number is a natural (a whole, non-negative) number greater than 1 that cannot be formed by multiplying two smaller natural numbers. While the notions of prime and composite numbers appear to have been first presented in the Egyptian Rhind Mathematical Papyrus from around 1550 BC (Bruins, 1974), the earliest surviving formal account of prime numbers comes from the Greek Euclid (around 300 BC). Euclid's Elements prove the infinitude of primes and the fundamental theorem of arithmetic, which states that every integer (a number that can be written without a fractional component) greater than 1 either is a prime number itself or can be represented as the product of prime numbers and that, moreover, this representation is unique, up to the order of the factors (Stillwell, 2010). Prime numbers have been later addressed by many notable mathematicians, including Dirichlet (e.g., Apostol, 1976), Euler (e.g., Sandifer, 2014), Fermat (e.g., Boklan and Conway, 2017), Gauss (e.g., Lenstra and Pomerance, 2002), Legandre (e.g., Chan, 1996) and Riemann (e.g., Apostol, 2000). There is, however, no general formula that generates all prime numbers.

The practical advantage of primality may be explained by the ability of different prime numbers to generate large numbers of different combinations without repetition. This property of prime numbers has been characterized in purely mathematical contexts (e.g., Furstenberg, 1955) and further in the geometric contexts of polygon construction and partition (Krizek *et al.*, 2001). Its applicability in digital computation has been noted in the contexts of hash table generation (Cormen *et al.*, 2001), error detection (Kirtland, 2001) and pseudo-random number generation (Matsumoto and Nishimura, 1998). Prime numbers have been shown to play fundamental roles in quantum physics (Peterson, 1999) and in quantum computation (Bengtsson and Zyczkowski, 2017) as well as the natural survival game of predator and prey (e.g., Williams and Simon, 1995), literature (e.g., Ribenboim, 2017) and music (e.g., du Sautoy, 2003). Yet, it appears that only very recently has a role been suggested for prime numbers in neuroscience (Baram, 2020a). As this role, namely, neural circuit categorization by polarization, is, as we have shown, central to information organization in the brain, prime numbers constitute an important part of this book.

1.2.11 *Quantum computation*

Twentieth century physics has put the scientific community on notice that the so-called classical way of information representation falls short of accommodating certain physical phenomena. The emerging quantum physics has led the way to quantum information representation, which, in turn, pushed the limits of quantum computation beyond previously assumed classical capacities. The introduction of Shor's algorithm for factoring numbers in polynomial time (Shor, 1994) has demonstrated the ability of quantum computation to solve certain problems more efficiently than classical computers. This perception was ratified two years later, when Grover (1996) introduced a quantum search algorithm which finds an item in a database in the square root of the time required by classical search. A comprehensive mathematical theory of quantum computation has matured (Nielsen and Chuang, 2000) and has been associated with cortical operation (Ivancevic and Ivancevic, 2010; Clark, 2014).

1.2.12 *Sensorimotor control*

Some of the more essential functions of the brain involve movement, sensation, and, in particular, a combination of the two. It is not surprising therefore, that major sections of the brain in animals and humans alike are devoted to the control of such composite functions. Perhaps the most fundamental notion underlying control theory is the contrast between open-loop and closed-loop control systems. While early clinical studies have implicitly employed closed-loop sensorimotor control (specifically, visual feedback in gait entrainment, Martin, 1967), the fundamental difference between open and closed-loop control in neurophysiological systems were only explicitly noted in later experimental studies. Specifically, in contrast to certain technological proposals for aiding movement in the neurologically impaired employing open-loop control by constantly moving visual cues (Riess and Weghorst, 1995; Hanakawa *et al.*, 1999), experimental (Kalaska *et al.*, 1989) and theoretical (Baram, 1999) studies have stressed the closed-loop effect of movement with respect to earth-stationary visual scenery, which is self-regulating (Seborg *et al.*, 2011). Similarly, closed-loop auditory gait entrainment (Baram and Miller, 2007) is self-regulating, unlike open-loop entrainment (Nieuwboer *et al.*, 2007).

1.3   Book outline

Following this introduction and historical account (Chapter 1) and some essentially mathematical preliminaries (Chapter 2), this book divides into four inter-related parts, which may, nevertheless, be independently read as separate entities. The first part, Cortical Graphs and Neural Circuit Primes (Chapters 3–6) introduces neural circuit structure as viewed from the perspectives of electrical somatic and synaptic polarities, graph theoretic connectivity and prime number interaction. The underlying premise, yielding a linguistic capacity exponential in the number of circuit neurons is that cortical connectivity, behavior and memory are all controlled by on/off gates, corresponding to neuronal somatic and synaptic membrane polarities with respect to a certain potential value. Graph theoretic considerations are shown to associate subcritical connectivity probability with small circuit segregation. These, in turn, are shown to resolve the implausible linguistics issue associated with large neural networks (Chapter 3). The resulting storage capacities of cortical information in terms of both circuit connectivity and firing-rate dynamics are demonstrated (Chapter 4). Primal-size categories of segregated circuit polarity code words are derived and argued to explain certain experimental results on "magical numbers" in working memory (Chapter 5). The high temporal information capacities of primal-size neural circuits in trees of meta-periodic interaction, producing linguistically plausible, non-repetitious polarity code words are argued and demonstrated (Chapter 6).

The second part of this book, Firing-rate Linguistics (Chapters 7–12), addresses the firing dynamics of neural circuits. We first note (Chapter 7) that the widely accepted, experimentally supported mathematical model of plasticity-modulated neural firing-rate implies a non-invertible and often chaotic process, concealing both past and future values. Instead of the actual values of the firing-rates, the singularities of the model, defining its dynamic modes, then become, together with the polarity (or connectivity) code, the language of cortical information. Hypothesizing that neural firing and plasticity time constant increase with age, discrete iteration maps corresponding to early development and maturity stages are then derived (Chapter 8) along with their singularities (Chapters 9 and 10). Neural circuits are shown to be segregated by neuronal polarization and elimination into smaller synchronous and asynchronous circuits (Chapter 11). Changes in synchronous circuit size, caused by changes in neuronal membrane polarities, are shown to change circuit firing-rate modes. Circuit segregation by synapse silencing is shown to yield interference-free asynchrony between synchronous subcircuits, facilitating, by different firing-rate dynamics, simultaneous execution of different cortical functions. Convergence of synaptic weights are shown to guarantee stable circuit memory; however, persistent modes of synaptic dynamics can also serve as a form of memory (Chapter 12). Following memory deterioration, caused by membrane silencing due to external inputs or electric charge dissipation, concealed memory, maintained by constant synaptic weights, is restored by presentation of the original circuit activation. Partial and false memory, as well as novelty and innovation effects, in the form of new circuit structures and firing dynamics, are created by incomplete deterioration or restoration of the circuit polarity which has instigated the original memory.

The third part of the book, Cortical Quantum Effects (Chapters 13 and 14), addresses the possible benefits of quantum search in such cortical functions as associative memory. Following a brief preliminary introduction to quantum computation and the Grover search algorithm (Chapter 13), the application of a quantum intersection set search procedure (Salman and Baram, 2012) to quantum pattern stability, completion and correction is presented along with capacity results and comparison to previous works. It is particularly noted that, much like the graph-theoretic subcritical probability results, advocating a small-circuit basis for high-performance, linguistically plausible cortical operation, the quantum-based results also advocate small circuits over large ones for such performance.

The fourth part of the book, Sensorimotor Control (Chapters 15–17), takes a more specific look at behavioral consequences of cortical activities, normal and impaired. In particular, cortical effects are considered in the context of sensorimotor control. Following consideration of normal sensorimotor behavior in human and bird (Chapter 15), clinical results demonstrate how such behavior is challenged by neurological impairment, and how neurologically impaired behavior can be improved by technological means (Chapter 16). Clinical results are used to demonstrate the hypothesized roles of memory and reward in sensorimotor control. We close by revisiting electroencephalogram (EEG) recordings (Chapter 17), ratifying the existence of positive and negative polarization in directional connectivity between functional cortical regions (specifically, occipital, parietal and motor regions), and singularity segregation, evidenced by simultaneous firing at different frequency (specifically, delta, alpha and beta) bands.

Chapter 2

Some Mathematical Preliminaries

2.1 Introduction

The mathematical disciplines involved in the analysis of neural circuits, as presented in this book, are quite diverse. Although these different disciplines will be shown to combine into a coherent theory of cortical connectivity and neuronal firing dynamics, they historically span several centuries of individual evolution, and did not necessarily share the same set of rules or logic. The present collection of mathematical preliminaries is not meant to replace an in-depth tutorial but rather to briefly familiarize the reader with the main concepts underlying this book, so as to facilitate a reasonably smooth progression through the subsequent chapters.

A key part of this book, namely the chapters addressing the nature of neural circuit polarity, criticality and primality, is independent of the specific mathematical and physical nature of neuronal firing and plasticity. It will equally apply to different dynamical models, such as the ones associated with spiking or firing-rate. Yet, in order to illustrate the different concepts of interest, we will specifically address the firing-rate model. Different models may require different biophysiological and mathematical elaborations, and may be interesting, even useful. Yet, addressing them all in one book would be an exceedingly demanding, highly sporadic task, and are therefore omitted in favor of a more coherent effort. The preliminaries presented here are all in the realm of classical mathematics. A separate section addressing quantum computation preliminaries is included in the third part of the book, considering quantum cortical effects.

2.2   Random graph criticality and connectivity

Graphs are mathematical structures which are widely used to model pairwise relationship between objects. The objects are referred to as *nodes* (or vertices), and the relationship between two nodes is reduced to their connectivity, which is represented by directional or bidirectional *edges*. *Random* graphs are characterized by the probability of their nodes being connected. As the context in which we employ the graph concept here is specific to neural circuits, we mention only certain results from *random graph theory* which are immediately relevant to neural circuits. More general definitions, assumptions and implications are avoided as much as possible. For further details and proofs of these results, the reader is referred to the cited references.

The number of nodes in a graph is denoted n. The probability of edge existence is denoted p. The number of edges is denoted m. For a given n, a graph is characterized by either m or p and denoted $G(n, m)$ or $G(n, p)$, respectively.

A graph is said to be *connected* if any pair of its nodes is connected by a sequence of edges.

Let

$$p = [\{\log n + c\}/n] \tag{2.1}$$

and

$$m = \left[\frac{n}{2}\log n + cn\right] \tag{2.2}$$

where c is a fixed real number quantifying the relationships of p and m to n and $[x]$ denotes the integer part of x, and let $P(G((n, p)))$ and $P(G(n, m))$ denote the probabilities of the corresponding graphs being connected. Then (Erdös and Rényi, 1959, 1960)

$$\lim_{n\to\infty} P(G(n, p)) = e^{-e^{-2c}} \tag{2.3}$$

and

$$\lim_{n \to \infty} P(G(n, m)) = e^{-e^{-2c}} \tag{2.4}$$

where e is the so-called natural logarithm approximated by $e = 2.7182$.

Both results imply convergence in probability to complete graph connectivity as $n \to \infty$.

A *connected component* of a graph is a connected subgraph none of whose nodes are connected to a node outside the subgraph.

It has been established (Erdös and Rényi, 1959, 1960) that, in the case of a subcritical graph, where $p = c/n$, with $0 < c < 1$, the size $S(n, p = c/n)$ of the largest connected component satisfies with probability tending to 1 as $n \to \infty$,

$$S(n, p = c/n) = O(\log(n)) \tag{2.5}$$

Detailed formal variants of Eq. (2.5) have been derived for homogeneous (Erdös and Rényi, 1959, 1960) and inhomogeneous (Bollobás *et al.*, 2007; Turova, 2010) random graphs ("homogeneous" refers to a common typical number (or "degree") of edges associated with different nodes).

In the case of a *supercritical* graph, where $p \geq (1 + \varepsilon)/n$ for any constant $\varepsilon > 0$, there is, with high probability, a single connected *giant component* whose size exceeds $O(\log(n))$, while all the other connected components have size $O(\log(n))$ (Erdös and Rényi, 1959, 1960). In the intermediate *critical* case $p = 1/n$, the size of the largest connected component of the graph is, with high probability, $O(n^{2/3})$ (Bollobás, 2001).

2.3 Primes, primality testing and prime factorization

A prime number (or "a prime") is a natural number greater than 1 that cannot be formed by multiplying two smaller natural numbers. A natural number greater than 1 that is not prime is called a composite number. For example, 5 is prime because the only ways of writing it as a product, 1×5 or 5×1, involve 5 itself. However, 6 is composite because it is the product of two numbers (2×3) that are both smaller than 6. Primes are central to number theory due to the fundamental theorem of arithmetic: every natural number greater than 1 is either a prime itself or can be factorized as a product of primes that is unique up to their order. The property of being prime is called primality.

Primality testing: A practical application of prime numbers would normally involve primality testing. A simple but slow method of checking the primality of a given number n, called trial division, tests whether n is a multiple of any integer between 2 and $\sqrt{n}$. Faster algorithms include the Miller–Rabin primality test (Miller, 1976; Rabin, 1980), which is often implemented in two stages: the first statistical (Rabin, 1980), the second deterministic, for verification (Miller, 1976). The AKS (Agrawal, Kayal and Saxena, 2004) primality test produces the correct answer in polynomial time, but is too slow to be practical. Particularly fast methods are available for numbers of special forms, such as Mersenne numbers, that is, prime numbers of the form $2^n - 1$ for some integers n. Finding those n values which satisfy Mersenne primality and those that do not has a long and interesting history, involving leading mathematicians (see, e.g., Laroche *et al.*, 2018).

Prime factorization of composite numbers: Given a composite integer n, the task of providing one or all prime factors of n is referred to as *prime factorization* of n. In terms of computational complexity, it has been established that prime factorization is in the complexity class NP (Nondeterministic Polynomial time, Pratt, 1975). It is, in general, significantly more difficult than prime testing. Trial division (Mollin, 2002) and Pollard's rho algorithm (Pollard, 1975) can be used to find very small factors of n (Riesel, 1994). Elliptic curve factorization can be effective when n has factors of moderate size (Lenstra, 1987). Methods suitable for arbitrary large numbers that do not depend on the size of its factors include the quadratic sieve and the general number field sieve (Pomerance, 1996). As with primality testing, there are also factorization algorithms that require their input to have a special form, including the special number field sieve (Pomerance, 1996).

Semiprime numbers: A *semiprime* is a natural number that is the product of two primes. The two primes in the product may equal each other, so the semiprimes include the squares of prime numbers. Because there are infinitely many primes, there are also infinitely many semiprimes. In 1974 the Arecibo message was sent with a radio signal aimed at a star cluster (Cassiday, 2013). It consisted of 1,679 binary digits intended to be interpreted as a 23×73 bitmap image. The number $1679 = 23 \times 73$ was chosen because it is a semiprime and therefore can be arranged into a rectangular image in only two distinct ways (23 rows and 73 columns, or 73 rows and 23 columns). The alternative arrangement, 23 rows by 73 columns, produces an unintelligible set of characters. The Arecibo message is noted here as an example of using prime factorization as means of defining a two-dimensional binary array. While our investigations of cortical information representation in this book do not directly involve two-dimensional arrays generated by prime factorization, they do give rise to such three-dimensional arrays, as described next.

Sphenic numbers: A sphenic number is a product pqr where p, q and r are three distinct prime numbers. This definition is more stringent than simply requiring the integer to have exactly three prime factors. For instance, $60 = 2^2 \times 3 \times 5$ has exactly three prime factors but is not sphenic. A literature check reveals purely mathematical interest in sphenic numbers (Lehmer, 1936). However, as we show in this book, sphenic numbers play a unique role in the context of neural circuits. Specifically, while circuits of 1–4 neurons have primal polarity code sizes, circuits of 5 and 6 neurons have polarity codes of sphenic sizes. This means that the corresponding codes can be factorized into primes only in three ways. This establishes a certain definition of the cortical language associated with such circuits.

2.4 Continuous and discrete-time dynamical systems

Poincare's attention in the late 19th century to low-dimension dynamical system has paved the way to new mathematics, culminating in the emergence of chaos theory during the second half of the 20th century. The notion of a limit cycle (Leloup *et al.*, 1999), and other "simple" attractors, has made way to that of "strange attractors". The revolutionary idea was that singularities, characterized by particular dynamic behaviors, are not necessarily points in state space but can also be orbits of different types and even whole domains in state space. The mathematical discipline of dynamical systems has become rather abstract in nature, making itself almost inaccessible to a non-mathematician. Yet, certain subcategories of dynamical system models are highly applicable in such domains as control systems, including neural circuits and networks. Such models are briefly presented below.

A first-order vector-valued differential equation over the reals, $v(t) \in R^n$, takes the form

$$\frac{dv(t)}{dt} = F(t, v(t)); \ v(0) = v_0 \tag{2.6}$$

where F is a vector-valued function said to be a dynamical system due to the dependence of the vector $v(t)$ on time t. An approximate solution of Eq. (2.6), known as Euler's method of discretization, is written as

$$v(k) = v(k-1) + \Delta F(t(k-1), v(k-1)) \tag{2.7}$$

where Δ is a "sufficiently short" distance along a line tangent to the curve $v(t)$ at $v(k-1)$, and $t(k) = t(k-1) + \Delta$. An equation of the form of Eq. (2.7) ia called a *discrete iteration map*. Clearly, the value of Δ is a key determinant of the accuracy of the method, which is often considered "sufficiently accurate" if it does not become numerically unstable, or "stiff" (Lambert, 1992).

We can be helped (as we shall be in the main part of this book, where we consider a cortical development stage termed *maturity*), when the system can, in a certain "piecewise" sense, be put in the form

$$\frac{dv(t)}{dt} = -Av(t) + N(v(t)) \tag{2.8}$$

where the first term on the right-hand side is linear and the second is non-linear in $v(t)$. Approximating the value of $N(v(t))$ by a constant,

$N(\boldsymbol{v}(k-1))$, over the interval $[t(k-1), t(k) = t(k-1) + \Delta]$, the discrete-time solution of Eq. (2.8) has the piecewise linear form

$$\boldsymbol{v}(k) = e^{-A\Delta}\boldsymbol{v}(k-1) + A^{-1}(1 - e^{-A\Delta})N(\boldsymbol{v}(k-1)) \qquad (2.9)$$

2.5 Elements of non-invertibility

A discrete iteration map over the reals $(\boldsymbol{v}(0), \boldsymbol{v}(k) \in R^n, k = 1, 2, \ldots)$

$$\boldsymbol{v}(k) = F(\boldsymbol{v}(k-1)); \quad \boldsymbol{v}(0) = \boldsymbol{v}_0 \qquad (2.10)$$

will be said to be *invertible* if the inverse map

$$\boldsymbol{v}(k-1) = F^{-1}(\boldsymbol{v}(k)) \qquad (2.11)$$

has a unique real solution. If this is not the case, Eq. (2.10) will be said to be a *non-invertible* map.

For instance, consider the scalar logistic map (May, 1976) with a parameter value 3

$$v(k) = 3v(k-1)(1 - v(k-1)) \qquad (2.12)$$

depicted in Fig. 2.1. It has its maximal value $v(k) = 0.75$ for $v(k-1) = 0.5$. It can be seen that the inverse map

$$v(k-1) = \frac{3 \pm \sqrt{9 - 12v(k)}}{6} \qquad (2.13)$$

has no real solution for $v(k) > 0.75$, a unique real solution $(v(k-1) = 0.5)$ for $v(k) = 0.75$, and a multiplicity (of two) real solutions for $0 \leq v(k) < 0.75$. In each of the subdomains $0 \leq v(k-1) \leq 0.5$ and $0.5 \leq v(k-1) \leq 1$, the inverse map, Eq. (2.13), does have a unique inverse. Such inverses have been called "partial inverses" (Abraham *et al.*, 1997). However, by the convention applied in this book, a map, representing a dynamical system, is *invertible* if and only if it is invertible everywhere in state space and everywhere in parameter space, and *non-invertible* if is non-invertible anywhere in state space or parameter space. The map, Eq. (2.12), is, then, non-invertible.

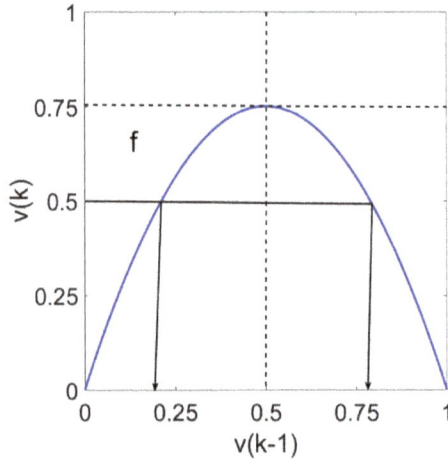

Figure 2.1.   The non-invertible discrete iteration map Eq. (2.12) has no inverse for $v(k) > 0.75$, a unique real inverse ($v(k-1) = 0.5$) for $v(k) = 0.75$, a multiplicity (two) of real solutions for $0 \leq v(k) < 0.75$, and partial inverses in each of the subdomains $0 \leq v(k-1) \leq 0.5$ and $0.5 \leq v(k-1) \leq 1$ (Abraham *et al.*, 1997).

2.6   Singularities of discrete iteration maps

It has been recognized that a meaningful characterization of non-linear dynamics consists in the identification of its singularities (Mira, 2007). A change in the nature of a singularity is called a *bifurcation*. The simplest singularities are period-k cycles (or periodic orbits) made up of k different consequent solution points of a discrete iteration map, satisfying $\boldsymbol{v}(n+k) = \boldsymbol{v}(n)$, $\boldsymbol{v}(n+r) \neq \boldsymbol{v}(n)$, $0 < r < k$. When $k = 1$, the point $\boldsymbol{v}(n)$ is called a *fixed point* (period one cycle). A k-cycle is *attracting* if all the eigenvalues of the Jacobian of the map F at all points of the cycle are smaller than 1 in magnitude, and *repelling* if any of these eigenvalues is greater than 1 in magnitude. Other types of attractors (e.g., *largely cyclic*) will be presented in subsequent sections, where we consider cortical firing-rate maps. Here, however, we note one more type of attractor, namely, a *chaotic* attractor, which, often referred to as a *strange* attractor, has several definitions. Although theoretical aspects of chaos, a centerpiece of dynamical systems theory for the past 30 years, are well understood, the empirical detection of chaos in a given time series is a non-trivial task. Mathematical measures, such as the largest Lyapunov exponent (Wright, 1984), often produce ambiguous empirical results for limited time series, even when applied to data generated by simulating low dimensional models (Sprott, 2003). Consequently, the intuitive characterization of chaos as a "deterministically unpredictable" process (Elyadi, 1999) often seems as reliable as any formal empirical measure. For practical analytic purposes, we resort to a formal definition of a chaotic map proposed by Li and Yorke (1975).

Given an interval J, a continuous map $f : J \to J$ is chaotic in the Li-Yorke sense if the following conditions are satisfied:

(1) For every $n = 1, 2, \ldots$ there is a periodic point in J having period n.

(2) There is an uncountable ("scrambled") set $S \subset J$, containing no periodic points, which satisfies the following conditions:

    (a) For every $x, y \in S$ with $x \neq y$

        (a1) $\displaystyle\limsup_{k \to \infty} \| f^k(x) - f^k(y) \| > 0$
        and
        (a2) $\displaystyle\liminf_{k \to \infty} \| f^k(x) - f^k(y) \| = 0$

    (b) For every $x \in S$ and any periodic point y of f,

$$\limsup_{k \to \infty} \| f^k(x) - f^k(y) \| > 0$$

where $f^k = f(f^{k-1}(x))$. Condition 2a implies sensitivity to initial conditions. While trajectories originating at different initial conditions will intersect each other (condition 2a1), they will not converge to each other (condition 2a2). Condition 2b implies that for a given subset (S) of initial conditions, the periodic points of the map are not asymptotically attractive.

The Li-Yorke theorem (1975) states that if there is a periodic point with period three (a 3-cycle), then conditions (1) and (2) above are satisfied, hence, the map f is chaotic.

2.7   Cobweb diagrams

Singularity analysis of continuous-time models has been the tradi-
tional framework for identifying stability and periodicity in system
dynamics (e.g., Arnold, 1989), adopted in the analysis of neural firing
dynamics (Cohen and Grossberg, 1983; Peterfreund and Baram,
1994a,b). However, the ability of discrete iteration maps to describe
particularly intricate dynamics in low dimensional models with
graphical clarity has made them highly popular in mathematical
circles for the past half century (e.g., Ott, 1993). Cobweb diagrams
(Koenigs, 1884; Lemeray, 1895; Knoebel, 1981; Abraham et al., 1997)
have been particularly useful in the analytic and graphical description of scalar dynamics. Yet, as noted in this book, synchronous
neural circuit and subcircuit firing, implied by the neuronal filtering
property, facilitates scalar analysis by cobweb diagrams, which are,
therefore, used extensively in this book.

In order to demonstrate the utility of a cobweb diagram on a
rather elementary level, we employ the generalized version of the
logistic map, Eq. (2.12), parameterized by r

$$v(k) = rv(k-1)(1 - v(k-1)) \qquad (2.14)$$

which is a common representation of a logistic process (e.g., Murray,
2002). The behavior of a dynamical system is known to be governed
by attractors, induced by points of intersection of the map with the
diagonal $v(k) = v(k-1)$ (Abraham et al., 1997). An attractor is
called global if its basin of attraction is the entire state space. It
has been shown (e.g., Murray, 1989) that, for $0 < v(0) < 1$, the
map Eq. (2.14) has a stable fixed point attractor at the origin for
$0 \leq r \leq 1$, a non-zero stable fixed point attractor for $0 < r < 3$, an
oscillatory attractor for $3 \leq r \leq 3.828$, and a chaotic attractor for
$r > 3.828$. The end points of such r domains are often called points
of *bifurcation* (Abraham et al., 1997).

A cobweb diagram is constructed in the space spanned by the
coordinates $v(k-1)$ and $v(k)$by connecting an initial point of the
map horizontally to a point on the diagonal $v(k) = v(k-1)$, then
vertically to a point on the map, then horizontally to a point on the

diagonal again, and so on. Figure 2.2 shows global attractors in the four regimes, corresponding, respectively, to (a) $r = 1$, (b) $r = 2$, (c) $r = 3$, (d) $r = 4$. In each of the four plots representing these attractors, the map f is represented by a solid blue-colored line, the diagonal $v(k) = v(k-1)$ is represented by a dashed green line, and the attractor, represented by a red x or a red line segment.

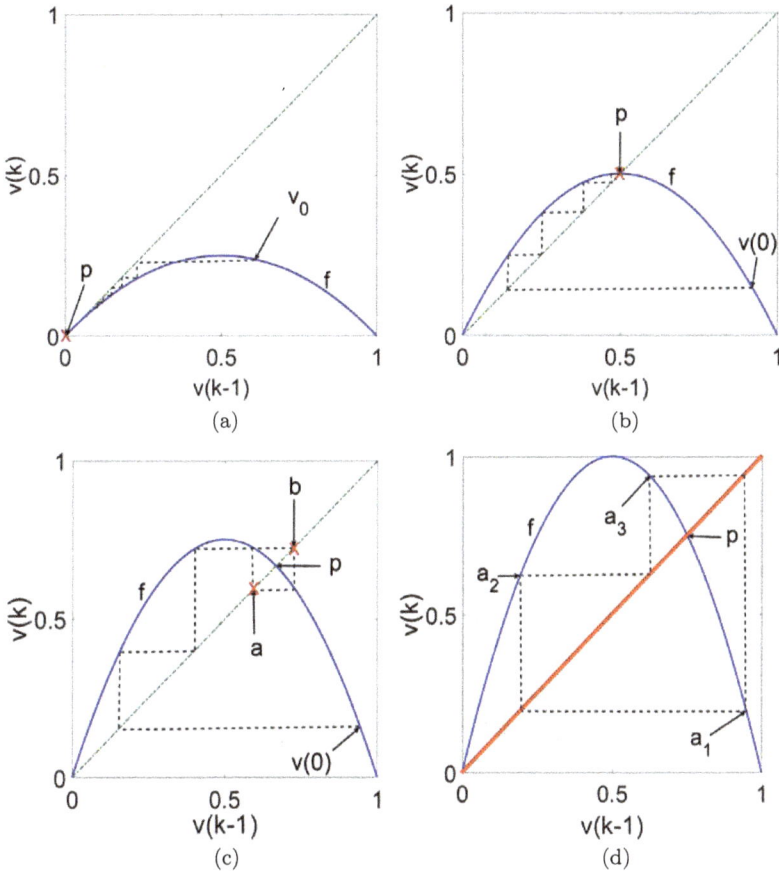

Figure 2.2. Attractor types of the logistic map Eq. (2.14): (a) silent for $r = 1$, (b) fixed-point for $r = 2$, (c) oscillatory for $r = 3$, and (d) chaotic for $r = 4$. The map f is represented by a blue line, cobweb trajectories, initiated at $v(0)$, are represented by black dashed line segment sequences, the diagonal $v(k) = v(k-1)$ is represented by a green dash-point line, and attractors are represented by red x or a red line segment.

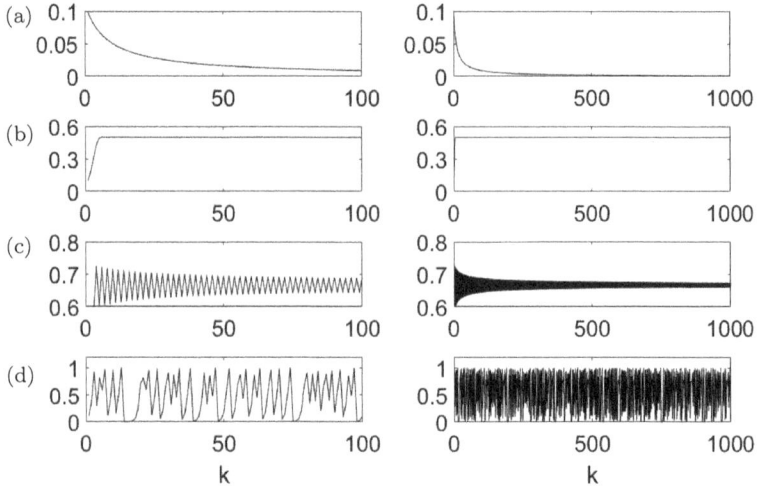

Figure 2.3. Simulated sequences of the logistic map, Eq. (2.14), for (a), (b), (c), (d). As predicted by the cobweb diagrams in Fig. 2.2, the sequences converge to (a) silent attractor, (b) fixed-point attractor, (c) oscillatory attractor and (d) chaotic attractor.

The attractor types represented by cobweb diagrams in Fig. 2.2 have been simulated for the same parameter (r) values, and are shown in Fig. 2.3.

While the rows (a)–(d) represent the four parameter values ($r = 1, 2, 3, 4$), the columns represent two different timescales (0–100 and 0–1,000), so as to illustrate the initial and convergent characteristics of the four firing modes. It can be seen that the four simulated cases indeed behave as might be expected from a silent (a), a fixed-point (b), an oscillatory (c) and a chaotic (d) attractor, respectively.

2.8   From continuous-time to discrete-time dynamics of neural firing and plasticity

The neuronal firing model has evolved from the integrate-and-fire (Lapicque, 1907) and the conductance-based membrane current (Hodgkin and Huxley, 1952) paradigms, through cortical averaging (Wilson and Cowan, 1972) and neuronal decoding (Abbott, 1994) to the spiking rate model (Gerstner, 1995).

$$\tau_{mi}\frac{d}{dt}\upsilon_i(t) = -\upsilon_i(t) + f_i\left(\boldsymbol{\omega}_i^T(t)\int_{-\infty}^{t}\mathbf{e}^{-(t-\sigma)/\tau_i}\boldsymbol{v}_i(\sigma)d\sigma + u_i(t)\right)$$

(2.15)

where $\upsilon_i(t)$ is the firing-rate of the i'th neuron, τ_{mi} is the membrane time constant, $\boldsymbol{v}_i(t)$ is the vector of firing rates of the circuit's pre-neurons of the i'th neuron, $\boldsymbol{\omega}_i$ is the corresponding vector of synaptic weights, τ_i is the synaptic time constant, $u_i(t)$ is the membrane activation potential, $u_i(t) = I_i(t) - r_i(t)$, with $I_i(t)$ an external input potential and $r_i(t)$ the membrane activation threshold (approximately -60 mV), and where

$$f_i(x) = f(x) = \begin{cases} x & \text{if } x \geq 0 \\ 0 & \text{if } x < 0 \end{cases}$$

(2.16)

is the conductance-based rectification kernel (Carandini and Ferster, 2000) first observed in empirical data (Granit *et al.*, 1963; Connor and Stevens, 1971).

The BCM plasticity rule (Bienenstock *et al.*, 1982) enhanced by stabilizing modifications (Intrator and Cooper, 1992; Cooper *et al.*, 2004) is a widely recognized, biologically plausible, mathematical representation of the Hebbian learning paradigm (Hebb, 1949). It suggests the computation of $\boldsymbol{\omega}_i(t)$ by the model

$$\tau_{\omega_i}\frac{d}{dt}\boldsymbol{\omega}_i(t) = -\boldsymbol{\omega}_i(t) + (\upsilon_i(t) - \theta_i(t))\boldsymbol{v}_i^2(t)$$

(2.17)

where τ_{ω_i} is the learning time constant and $\theta_i(t)$ is a variable threshold satisfying (Intrator and Cooper, 1992; Cooper *et al.*, 2004)

$$\theta_i(t) = \frac{1}{\tau_{\theta_i}}\int_{-\infty}^{t}\upsilon^2(\tau)\exp(-(t-\tau)/\tau_{\theta_i})d\tau$$

(2.18)

where τ_{θ_i} is the thresholding time constant.

Time discretization of Eq. (2.15) yields

$$\tau_{m_i} v_i(k+1) = (\tau_{m_i} - 1)v_i(k)$$
$$+ f_i\left(\omega_i(k)^T \sum_{\ell=-\infty}^{k} e^{-(k-\ell)/\tau_i} \boldsymbol{v}(\ell) + u_i(k)\right) \quad (2.19)$$

A change of variables $p = k - \ell$ yields

$$\tau_{m_i} v_i(k+1) = (\tau_{v_i} - 1)v_i(k)$$
$$+ f_i\left(\omega_i(k)^T \sum_{p=0}^{\infty} e^{-p/\tau_i} \boldsymbol{v}_i(k-p) + u_i(k)\right) \quad (2.20)$$

Since v_i can be assumed to be bounded, and since the error incurred by truncating the infinite sum

$$r_i \equiv \sum_{p=0}^{\infty} e^{-p/\tau_i} - \sum_{p=0}^{N} e^{-p/\tau_i} = \frac{e^{-N/\tau_i}}{1 - e^{-1}} \quad (2.21)$$

decays exponentially, the firing-rate model can be approximated to any desired accuracy by the model

$$v_i(k+1) = \frac{(\tau_{v_i} - 1)}{\tau_{v_i}} v_i(k)$$
$$+ \frac{1}{\tau_{v_i}} f_i\left(\omega_i(k)^T \sum_{p=0}^{N-1} e^{-p/\tau_i} \boldsymbol{v}(k-p) + u_i(k)\right) \quad (2.22)$$

for some integer N. The value of N may be selected as one which yields a sufficiently small value for the relative error

$$\rho_i = \frac{r_i}{\sum_{p=0}^{\infty} e^{-p/\tau_i}} = e^{-N/\tau_i} \quad (2.23)$$

Time discretization of Eqs. (2.17) and (2.18) yields

$$w(k) = \exp(-1/\tau_w)w(k-1) + (1 - \exp(-1/\tau_w))$$
$$\times (v(k-1) - \theta(k-1))v^2(k-1) \quad (2.24)$$

and

$$\theta(k) = 1/\tau_\theta \sum_{i=0}^{N} \exp(-i/\tau_\theta)v^2(k-i) \qquad (2.25)$$

2.9 Self-regulation and stability in open and closed-loop control systems

The fundamental difference between open-loop and closed-loop control systems is one of the more basic notions underlying control theory and its practice. Block diagrams representing the two system types are depicted in Fig. 2.4. It can be seen that, in the open-loop system (Fig. 2.4(a)), the input signal u drives a forward controller C, which, being in a state x, drives a plant P, producing an output signal y. On the other hand, in the closed-loop system (Fig. 2.4(b)), having the same forward-path as the open-loop system, the output signal y is measured by a sensor S, whose output s is fed into a feedback controller F, whose output f is subtracted from the input signal u.

For the purposes of this book (specifically, those associated with sensorimotor control, addressed in the latter part of the book), we consider two behavioral properties which are fundamental to the proper operation of a control system: stability and self-regulation. There are several mathematical definitions of control system stability,

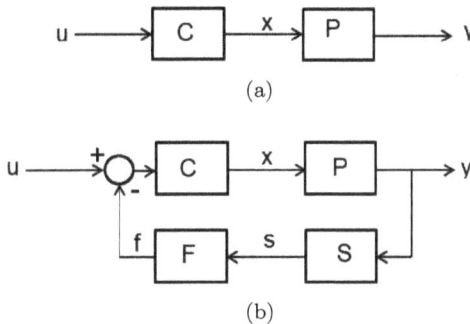

Figure 2.4. (a) Open-loop control system (b) closed-loop control system. With plant P, feedforward controller C, sensor S and feedback controller F.

of which the so-called "bounded-input-bounded-output" (BIBO) stability is immediately relevant. It means that there exists a constant k such that for all t_0 the statements $x(t_0) = 0$, $\|u(t)\| \leq 1$ and $t \geq t_0$ imply $\|y(t)\| \leq k$ for $t \geq t_0$ (e.g., Brockett, 1970). It can be shown by linearization and classical control theoretic arguments that both open-loop and closed-loop systems can be BIBO stable. Specifically, either type of system is stable if the poles of the corresponding transfer function under the Laplace transform are negative (e.g., Kuo, 1981).

The self-regulation property (e.g., Seborg *et al.*, 2011) implies that the output follows the input. Since, even under stability, the output of a controlled system may be externally disturbed and, thus, accumulate error with respect to the desired output, self-regulation is essential for proper control operation. Clearly, by virtue of feedback, only closed-loop systems can be self-regulated.

2.10   Discussion

The singularities of a map are often referred to as attractors (keeping in mind that the term "attractors" is also used for repelling singularities), induced by points of intersection of the map with the diagonal $v(k) = v(k-1)$ (Abraham *et al.*, 1997). An attractor is called global if its basin of attraction is the entire state space. The domain of a map generally decomposes into the basins of point attractors, cyclic attractors, chaotic attractors and an "attractor of infinity", which consists of all points whose trajectories run away from any bounded set (Abraham *et al.*, 1997). While mathematically, indefinite divergence of the neuronal firing-rates, to be discussed later in this book, is possible, physical constraints exclude such divergence. At the same time, asymptotic convergence to a periodic point from the scrambled set S is excluded by the Li-Yorke conditions (specifically, (2b)). Initiated within the scrambled set, a trajectory will enter the basin of a chaotic attractor, which, for a polynomial map, has a highly patterned geometric and statistical ("intensity") structure (Field and Golubitsky, 2009). (As we shall see, the cortical firing-rate map is also of a polynomial nature). Global bifurcations from chaotic and repelling basins (the latter a part of the "basin of infinity") into cyclic (and point) attractors are made possible by "absorbing areas" (Abraham *et al.*, 1997). A chaotic attractor may be viewed, then, as an integral part of the global attractor repertoire. However, cortical functionality appears to require separation of dynamic modes. In this book we suggest that mixing, drifting, absorption or bifurcation of dynamic modes are avoided by cortical circuit segregation, which, as we show, is the outcome of an elementary cortical property, namely, neural circuit polarity.

Part I
Cortical Graphs and Neural Circuit Primes

Chapter 3

Polarity Codes and Subcritical Linguistics

3.1 Introduction

Any representation of information involves a language. Natural languages, such as English, consist of elementary alphabets, comprising letters, which connect into words. Words, being the smallest embodiment of meaning, are normally short, consisting of two to about ten letters. Numbering in the millions and normally grouped into sentences, paragraphs, sections, chapters, articles and books, words carry the entire burden of linguistic information. Without such structures, information would be difficult to comprehend or memorize. Computer languages have conceptually similar structures. Their lowest-level alphabet consists of $(0, 1)$ bits. Computer words normally consist of 2–32 bits. Yet, computer programs, consisting of "commands" can be arbitrarily long. Both natural and artificial manifestations of information extend beyond the formal linguistic domain, employing vision, hearing, smell, touch and motion in the generation and memorization of information. Highly inspiring progress has been made in the formal conceptualization of cognitive functions, including behavioral decision making (Wei *et al.*, 2017), communicational behaviour linkage to neural dynamics (Bonzon, 2017), feeling of understanding (Mizraji and Lin, 2017), multisensory learning (Rao, 2018) and bilingual language control (Tong *et al.*, 2019).

A connected neural circuit represents a single cortical word. For a large neural circuit, this would imply a long code word, comprising binary polarity states, which not only consumes many neurons, but is difficult to comprehend linguistically. Neural circuit segregation into smaller circuits is, then, necessary for reasonably short code words in cortical language.

As we show in this chapter, the sizes of such circuits are implied by graph theoretic considerations. These are also found to govern the linguistics of learning and memory. Employing a polarity-based version of the so-called Hebbian paradigm, we show that linguistic characteristics, such as the lengths of cortical code words, are strongly dependent on the probability of connectivity, which is the probability of positive polarity.

It is widely accepted that the neural consistency of the brain is highly heterogeneous (e.g., Hu *et al.*, 2013; Baroni and Mazzoni, 2014; Han *et al.*, 2018). This often represents a variety of specifically defined neuronal types and neural circuit structures. As our main purpose here is to present a new graph-theoretic approach to the memory capacity of neural circuits, contrasting earlier information-theoretic results (McEliece *et al.*, 1987; Amit *et al.*, 1987; Baram and Sal'ee, 1992), which pertain to a particular, largely simplified and unified neural network model (McCulloch and Pitts, 1943; Amari, 1972; Hopfield, 1982), it seems appropriate to limit the discussion to the same model for comparison purposes.

On a conceptual level, the binary nature of neural networks in those information-theoretic studies is replaced here by the more recently discovered molecularly and physiologically-based notions of electrical membrane (Melnick, 1994) and synapse (Atwood and Wojtowicz, 1999) potentials being above or below a certain value (about -60 mV), which, defining inter-neural connectivity, have been called cortical circuit polarities (Baram, 2018). The diversion from the original information-theoretic models is represented by the graph-theoretic based notion of the connection probability criticality. The resulting segregated neural circuits constitute a form of cortical linguistics. Following our more recent publication (Baram, 2020b), we start by addressing the neuronal connectivity codes implied by somatic and synaptic polarities and their projection on circuit segregation suggested by random graph theory (Erdos and Renyi,

1959, 1960). A comparison of the information capacities under subcritical and supercritical neural circuit connectivity probabilities yields an unequivocal conclusion regarding the linguistic advantage of subcritically connected neural circuits.

3.2 Neuronal polarity codes

Neuronal membrane and synapse polarities can be described as on/off gates which, in the "off" ("disconnect") position, represent negative polarity, and, in the "on" ("connect") position, represent positive polarity. The binary polarity states of a neural circuit will represent a code word. The polarity gates of a single neuron are represented graphically in Fig. 3.1(a) by angular discontinuities in line segments representing membranes and synapses as specified. Neuronal self-feedback, employing an intermediate synapse between axonal output and neuronal input, has been suggested (Groves *et al.*, 1975), although this is not a generic concept and there are other self-feedback models (such as the axonal discharge model (Carlsson and Lindquist, 1963; Smith and Jahr, 2002). As in the rest of this book, our intent is to address certain concepts, rather than attempt to provide a complete account of the highly heterogeneous neural consistency of brain. As the neuron-external inputs directly affect the membrane polarity, they are accounted for in the 3-word polarity code depicted in Fig. 3.1(b)–(d) as: (b) positive membrane and self-synapse polarities, (c) positive membrane polarity and negative self-synapse polarity and (d) neuronal silence, implied by negative membrane and self-synapse polarities. Figure 3.2 shows a 2-neuron circuit, with all somatic and synaptic polarities in the negative ("off") state revealed for illustration purposes.

In order to derive the n-neuron circuit polarity code it is first noted that there can be up to n active neurons (neurons with positive membrane polarities) and n^2 synapses (including self-synapses). A synapse can be either active or silent. It follows that there can be up to

$$R^{(1)}(n) = 2^{n^2} \tag{3.1}$$

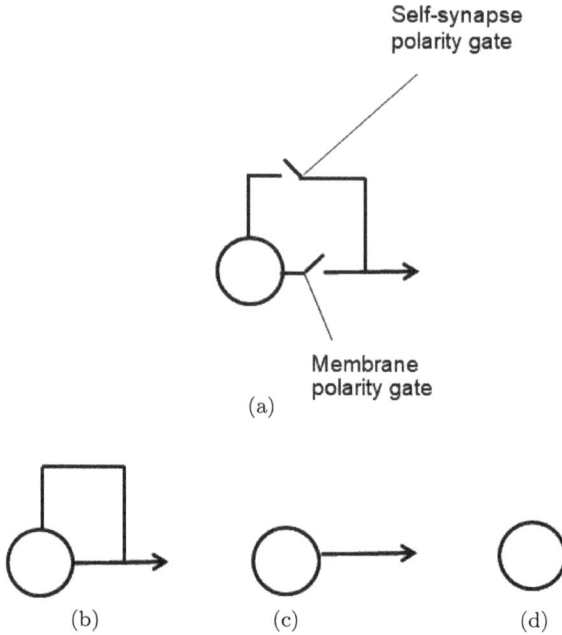

Figure 3.1. Neuronal polarity gates (a) and their code (b)–(d). As the neuron-external inputs directly affect the membrane polarity, they are accounted for in the neuronal 3-words polarity code, which specifies the connected elements only, without the "off" gates (Baram, 2020b).

different circuits involving only active membranes and active or silent synapses. In addition, there can be circuits with $i = 1, \ldots, n$ silent membranes, yielding

$$R^{(2)}(n) = \sum_{i=1}^{n} \binom{n}{i} 2^{(n-i)^2} \tag{3.2}$$

different polarity patterns. It follows that there is a maximum total of

$$R(n) = R^{(1)}(n) + R^{(2)}(n) = 2^{n^2} + \sum_{i=1}^{n} \binom{n}{i} 2^{(n-i)^2}$$

$$= \sum_{i=0}^{n} \binom{n}{i} 2^{(n-i)^2} \tag{3.3}$$

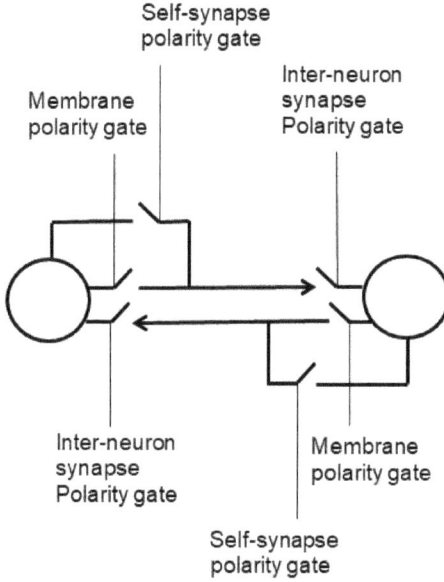

Figure 3.2. Two-neuron polarity-gated circuit. All polarity gates are open for graphical clarity (Baram, 2020b).

different circuits, constituting the size of the circuit polarity code of n polarity-gated neurons.

Equation (3.3) yields $R(1) = 3$, $R(2) = 21$, $R(3) = 567$, $R(4) = 67,689$, $R(5) = 33,887,403$, $R(6) = O(10^{10})$, $R(7) = O(10^{14})$, $R(8) = O(10^{19})$, $R(9) = O(10^{24})$, $R(10) = O(10^{30})$. The different circuits, constituting the size of the circuit polarity code of n polarity-gated neurons (Eq. (3.3) corrects an error in Baram, 2018, Eq. (3.2)). The polarity code for a circuit of two neurons is illustrated in Fig. 3.3, where "off" (negative) polarity gates are omitted for graphical clarity.

3.3 Subcritical neural circuit segregation

Cortical activity segregation and integration have been argued on grounds of thalamocortical simulations (Stratton and Wiles, 2015). A mathematical foundation for the relationship between polarity,

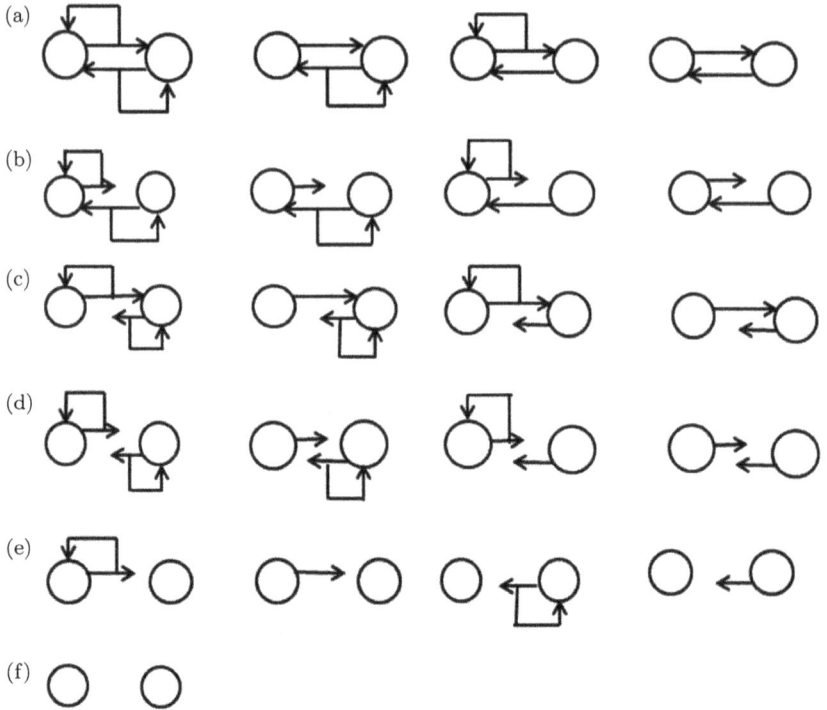

Figure 3.3. Polarity code of 2-neuron circuits. Each row represents one inter-neuron connectivity with all possible states of self-feedback. (a) Bidirectional interneuron connectivity (b) both neurons active with directional connectivity from right to left neuron (c) both neurons active with directional connectivity from left to right neuron (d) two segregated active neurons (e) one active neuron (f) two silent neurons (Baram, 2020b).

segregation and firing dynamics has been established (Baram, 2018). Synaptic polarization may clearly result in circuit segregation.

Graph-theoretic considerations (Erdos and Rényi, 1959, 1960) imply that, if the connectivity probability $p(n)$ between pairs of neurons in an assembly of n neurons with $n \to \infty$ satisfies the *subcriticality* condition $p(n) = c/n$ for fixed $c < 1$, then, with high probability, the largest connected circuit will be of maximal size $\log(n)$. On the other hand, if $p(n)$ satisfies the *supercriticality* condition $p(n) = c/n$ for fixed $c > 1$, then, with high probability, the largest connected circuit will be of maximal size linear in n. Finally,

(a)

(b)

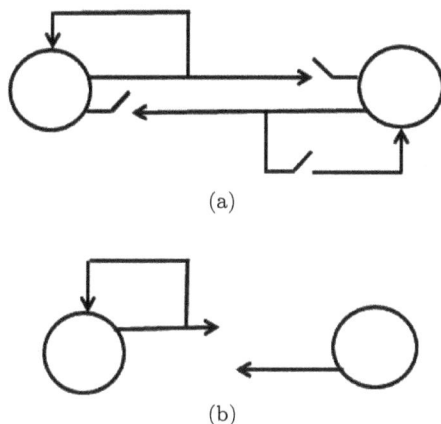

Figure 3.4. Given (a) a 2-neuron circuit with all polarity gates in the "off" states revealed. The mixed polarity pattern yields (b) two segregated 1-neuron circuits in different polarity states, with the "off" states expressed by disconnect (Baram, 2020b).

if $p(n)$ satisfies the *criticality* condition $p(n) = 1/n$, then the largest connected circuit will be of maximal size $n^{2/3}$ (Bollobás, 1984).

For reasons that will become apparent in the subsequent sections, we relax the above formal definitions somewhat by leaving the criticality condition $p(n) = 1/n$ as is, changing its notation to $p(n) \equiv q(n) = 1/n$ and defining subcriticality as the condition $p(n) < q(n)$ and supercriticality as the condition $p(n) > q(n)$.

Figure 3.4(a) displays a 2-neuron circuit with a certain polarity pattern. As shown in Fig. 3.4(b), the mixed polarity of Fig. 3.4(a) yields two mutually segregated 1-neuron circuits, with the polarity gates in the "off" states concealed for simplicity.

Noting that an average of 7×10^3 synapses per neuron (Drachman, 2005) in the average human brain of $n = 10^{11}$ neurons (von Bartheld *et al.*, 2016) would yield an average connectivity probability of

$$p(n) = 7000/10^{22} < 10^{-18} = 10^{-7} \times 10^{-11} = 10^{-7}n \qquad (3.4)$$

which satisfies the sub-criticality condition $p = c/n$, $0 < c < 1$. This would imply an upper bound of $\log(n)$ on the neural circuit size

(Eq. (2.5), Chapter 2), which, in turn, implies a lower bound

$$C(n) = \frac{n}{\log(n)} \qquad (3.5)$$

on the number of externally disconnected, internally connected neural circuits. In a subsequent chapter we address more specific benefits of subcriticality associated with cyclic information generation. Equation (3.5) constitutes, then, a lower bound on the average information capacity (in number of code words) of n subcritically connected neurons.

The number n may take different values corresponding to the sizes of different cortical structures, hence, different maximal segregated circuit sizes $\log(n)$. For instance, for $n = 1000$ we have $\log(n) = O(7)$ (more precisely 6.91), for $n = 25 \times 10^9$ (which is about the size of the human cerebral cortex) we have $\log(n) = O(24)$ and for $n = 10^{11}$ (which is about the size of the human brain) we have $\log(n) = O(25)$. As we have noted, the manipulation and memory of such long words (e.g. 24, 25 letters) is not normally attempted in natural languages and seems equally unrealistic in cortical terms.

3.4 Hebbian linguistic impasse of probabilistically supercritical neural polarity

The linguistic difficulties associated with the memorization and manipulation of brain-size code words in cortical language does not appear to have been explicitly noted in early studies of cortical information coding. In fact, experimental observations of simultaneous activity in large cortical areas have seemed to justify a large network approach to cortical information processing. Seminal studies on networks of binary neurons (McCulloch and Pitts, 1943; Amari, 1972; Hopfield, 1982) were followed by studies of their memory capacity. Mathematical manifestations of the Hebbian learning paradigm have produced a variety of bounds on the memory capacity of such networks (e.g., McEliece *et al.*, 1987; Amit *et al.*, 1987; Baram and Sal'ee, 1992). As these bounds increase with the number of neurons in the network, so do the lengths of the code words. While the first (McEliece *et al.*, 1987) and the third (Baram and Sal'ee, 1992) studies assumed that the network size n satisfies $n \to \infty$, the second study (Amit *et al.*, 1987) assumed a finite network size n, which is, however, allowed to grow indefinitely. Below, we consider all three models, so as to show that the linguistic impasse is inherent in the three models. In order to put these earlier binary models in the same context as the polarity models considered in the present work, let us assume that all synapses are in positive polarity (1-valued) state, while the membrane polarities of the neurons are in positive (1-valued) state with probability p and in negative (0 or -1-valued) polarity state otherwise.

In the first model (McEliece *et al.*, 1987), Hebbian storage is represented by the sum of outer products

$$W = \sum_{i}^{M} \omega^{(i)} \omega^{(i)T} \tag{3.6}$$

where $\omega^{(i)}$, $i = 1, \ldots, M$ are n-dimensional vectors of equiprobable ± 1 network polarities. Retrieval is performed according to

$$x_k = \begin{cases} 0, & \text{if } u_k \leq 0 \\ 1, & \text{if } u_k > 0 \end{cases} \tag{3.7}$$

where

$$u_k = \sum_{j=1}^{n} \omega_{k,j} x_j \tag{3.8}$$

with $\omega_{k,j}$ the corresponding element of the matrix W. An upper bound on the number of stored patterns

$$M < \frac{n}{2\log(n)} \tag{3.9}$$

yields equilibrium

$$W\omega^{(i)} = \omega^{(i)} \tag{3.10}$$

at each of the stored vectors $\omega^{(i)}$, $i = 1, \ldots, M$. Storage to full capacity would yield the inter-neuron connectivity probability

$$P(n) = 1 - (1 - p(n))^M \tag{3.11}$$

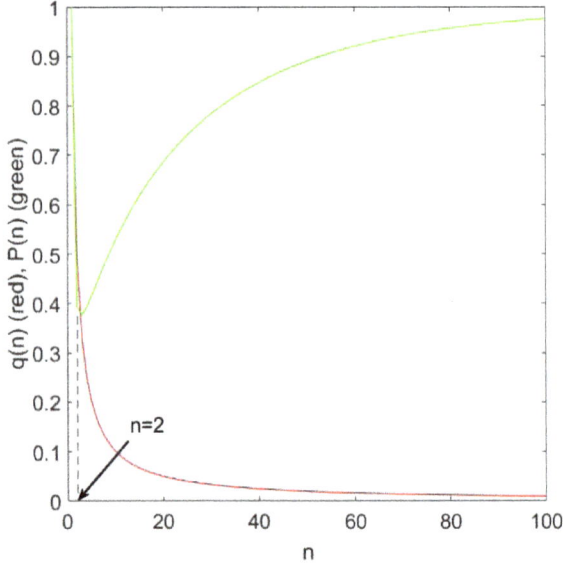

Figure 3.5. For polarity probability $p(n) = 0.5$ and the storage capacity bound specified by Eq. (3.9), the inter-neuron connectivity probability $P(n)$ (green curve), specified by Eq. (3.12), is ploted along with the critical connectivity probablilty $q(n) = 1/n$ (red curve). Beyond the critical (intersection) point at $n = 2$, $P(n)$ quickly converges to 1, which, implying large circuit connectivity, is increasingly unacceptable for linguistic purposes (Baram, 2020b).

which, under Eq. (3.9). yields

$$P(n) = 1 - (1 - p(n))^{n/2 \log(n)} \qquad (3.12)$$

$P(n)$ now takes the place of $p(n)$ as the neural network's probabilistic criticality measure, following the storage of M polarity patterns. Specifically, $P(n)$ is critical if $P(n) = q(n) = 1/n$, subcritical if $P(n) < q(n)$ and supercritical if $P(n) > q(n)$. It can be readily verified that, since $p(n) = 0.5$, $P(n)$ is supercritical for $n \geq 2$. The value of $P(n)$ as specified by Eq. (3.12) is plotted (green colored), along with the critical value $q(n) = 1/n$ (red-colored), against n in Fig. 3.5. The intersection point $P(n) = q(n)$ is at $n = 2$. It can be seen that as n approaches the value 100, $P(n)$ approaches the value 1, and becomes closer to 1 as n grows larger. This means that the circuit becomes, with high probability, connected, with maximal size linear in n (in fact, specifically n), which is consistent with the graph-theoretic property of supercriticality, and which, as we have noted, may be of little linguistic use, if any.

In the second model (Amit *et al.*, 1987), Hebbian storage is represented by the attenuated sum of outer products

$$W = \frac{1}{n} \sum_{i}^{M} \omega^{(i)} \omega^{(i)T} \qquad (3.13)$$

where $\omega^{(i)}$, $i = 1, \ldots, M$ are n-dimensional vectors of equiprobable ± 1 network polarities. Employing the physical spin-glass model (Hopfield, 1982), equilibrium stability was found to be destroyed beyond the following upper bound on the number of stored patterns

$$M < 0.14n \qquad (3.14)$$

(Amit *et al.*, 1987). While, in contrast to the other models noted (McEliece *et al.*, 1987; Baram and Sal'ee, 1992), the number of neurons, n, assumed in the present model (Amit *et al.*, 1987) is finite, the supercriticality effect is similar, as we show next. The inter-neuron connectivity probability, Eq. (3.11), takes, in the present case, the form

$$P(n) = 1 - (1 - p(n))^{0.14n} \qquad (3.15)$$

The value of $P(n)$ according to Eq. (3.15) is plotted (green colored) in Fig. 3.6, along with the critical value $q(n) = 1/n$ (red-colored). The critical (intersection) point $P(n) = q(n)$ is at $n = 3$. It can be seen that as n approaches the value 100, $P(n)$ quickly approaches the value 1, which, implying large circuit connectivity, is increasingly unacceptable for linguistic purposes.

In the third model (Baram and Sal'ee, 1992), as in the first model, the synaptic weight connecting the k'th and the j' th neurons is obtained from a set of vectors over $\omega^{(i)} \in \{0, 1\}^n$, $i = 1, \ldots, M$. Storage is performed by the modified Hebbian rule (Tsodyks and

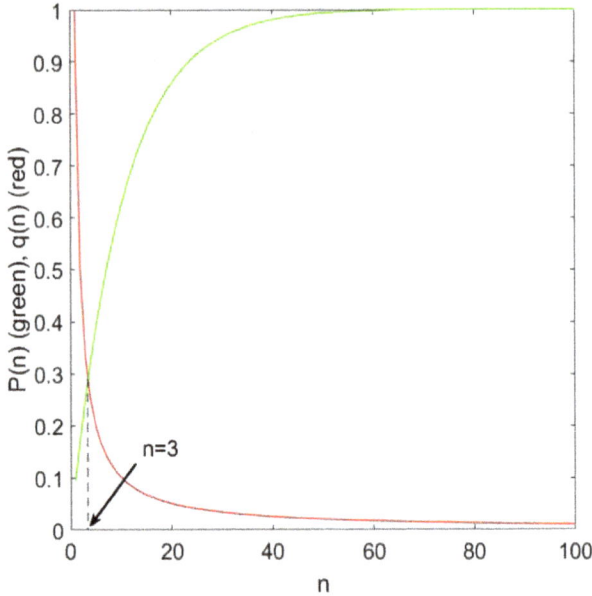

Figure 3.6. For supercritical polarity probability $p(n) = 0.5$ and the storage capacity specified by Eq. (3.14), the inter-neuron connectivity probability $P(n)$ (green curve), specified by Eq. (3.15), is ploted along with the critical connectivity probablilty $q(n) = 1/n$ (red curve). Beyond the critical (intersection) point at $n = 3$, $P(n)$ quickly converges to 1, which, implying large circuit connectivity, is increasingly unacceptable for linguistic purposes (Baram, 2020b).

Feigel'man, 1988; Vincente and Amit, 1989)

$$\omega_{k,j} = \sum_{i}^{M} (\omega_k^{(i)} - p)(\omega_j^{(i)} - p) \qquad (3.16)$$

where p is the probability of a stored membrane polarity having the value 1.

The recalled polarity states x_k, $k = 1, \ldots, n$ are obtained by the McCulloch-Pitts rule

$$x_k = \begin{cases} 0, & \text{if } u_k \le t \\ 1, & \text{if } u_k > t \end{cases} \qquad (3.17)$$

where

$$u_k = \sum_{j=1}^{n} \omega_{k,j} x_j \qquad (3.18)$$

and $t = \alpha n p$, with $\alpha = (4 - \tau)/4(2 - \tau)$, is a threshold.

Let us denote the probability that any of the stored polarity vectors is not an equilibrium point P_e, and, further, let $p = p(n) = cn^{-\tau}$ with $0 < c \leq 0.5$ and $0 < \tau < 1$. Then $P_e \to 0$ as $n \to \infty$ if

$$M \leq \gamma \frac{n}{p(n) \log n} \qquad (3.19)$$

for any $\gamma < 1/4(2 - \tau)$. Eq. (3.19) represents, then, an upper bound on the equilibrium polarity capacity of a neural circuit (Baram & Sal'ee, 1992, Theorem 1). While the latter study also addressed the error correction capacity of the network, here we restrict the analysis to equilibrium capacity for simplicity.

Clearly, for any $0 < c \leq 0.5$ and $0 < \tau < 1$, $p(n) = cn^{-\tau}$, the smaller the value of c (say, $c = 0.01$) and the larger the value of τ (say, $\tau = 0.99$), the smaller the value of $p(n)$ and the higher the bound in Eq. (3.19). Storage to full capacity would yield inter-neuron storage connectivity probability which, under Eq. (3.19), yields

$$P(n) = 1 - (1 - p(n))^{\gamma n/p(n) \log(n)} \qquad (3.20)$$

It can be readily verified that for $n > 1$, $P(n) > 0.5$, hence, $P(n)$ is supercritical. The value of $P(n)$ as specified by Eq. (3.20), where $p(n) = cn^{-\tau}$ with $c = 0.01$ and $\tau = 0.99$, is plotted (green colored), along with the *critical* value $q(n) = 1/n$ (red-colored), against n in Fig. 3.7. The intersection point $P(n) = q(n)$ is at $n = 1$. It can be seen that, as n approaches the value 100, $P(n)$ approaches the value 1. This means that the circuit becomes connected with increasingly high probability, which, as we have noted, may be of little linguistic use, if any. It might be noted that, since the shapes of the curves depicted in Figs. 3.5 to 3.7 are similar, the language of their corresponding verbal descriptions is also similar. The distinction between these figures is specifically expressed by the corresponding

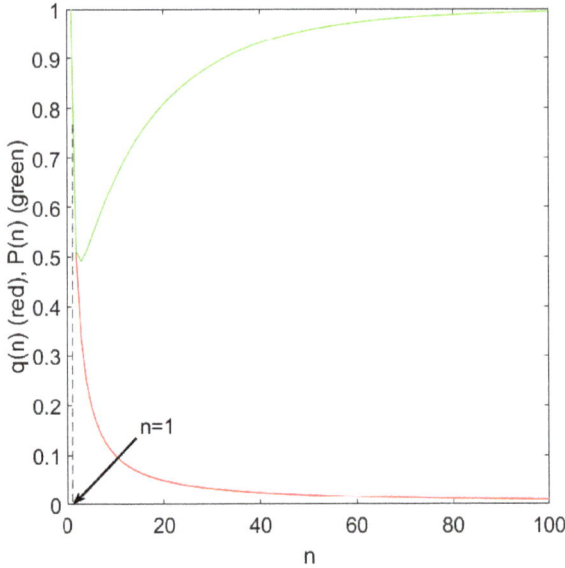

Figure 3.7. For polarity probability $p(n) = 0.01n^{-0.99}$ the inter-neuron storage connectivity probability is $P(n)$ (green curve) and the critical connectivity probablilty is $q(n)$ (red curve). Beyond the critical (intersection) point at $n = 1$, $P(n)$, which is supercritical, quickly converges to 1, which, implying large circuit connectivity, is increasingly unacceptable for linguistic purposes (Baram, 2020b).

Eqs. (3.12), (3.15) and (3.20), which, while yielding similar results, are clearly different.

3.5 Critical Hebbian linguistic limit of neural polarity

Of the three models considered above, only the third (Baram and Sal'ee, 1992) allows for a variable polarity probability. Let us now assume membrane polarity probability $p(n) = c/n$ with $0 < c \leq 0.5$. The linguistic capacity will be controlled by the parameter c, which is characteristically dependent on the size and the functionality of the corresponding cortical domain. Subcritical polarity implies that the largest connected circuit (having a sequence of connections between any two neurons) is of size $\log(n)$ (Erdos and Renyi, 1959, 1960).

As, for large n, a subcritical probability $p(n)$ implies a highly sparse connectivity, the storage of

$$M = \log(n) \tag{3.21}$$

circuit polarity vectors according to Eq. (3.6) implies that these vectors are, with high probability, mutually orthogonal and are therefore, equilibrium points of the circuit satisfying

$$W\omega^{(i)} = \omega^{(i)} \tag{3.22}$$

where W is the matrix whose elements, $\omega_{k,j}$, are defined by Eq. (3.6), where $\omega^{(i)}, i = 1, \ldots, M$ are the n-dimensional polarity vectors stored and $M = O(\log(n))$.

Employing Eqs. (3.5) and (3.21), the Hebbian equilibrium storage capacity for the entire assembly of n neurons is

$$B(n) = O(C(n)\log(n)) = O(n) \tag{3.23}$$

where $C(n)$ is defined by Eq. (3.5). From a linguistic viewpoint, the significance of circuit segregation is that each of the words is of maximal length $\log(n)$, which is, for large n, considerably smaller than n.

The storage inter-neuron connectivity probability is, by Eq. (3.21)

$$P(n) = 1 - (1 - p(n))^{\log(n)} \tag{3.24}$$

For $c = 0.5$, we have the polarity probability $p(n) = 0.5/n$. In Fig. 3.8, the critical function $q(n)$ is represented by the red curve and the inter-neuron storage connectivity probability, $P(n)$, is represented by the green curve. In the subcritical domain of $P(n)$, the maximal connected circuit size is $s = [\log(7)] = 2$ (with $[x]$ representing the integer closest to x), which is the critical linguistic limit of the neural polarity code word size limit. For $n > 7$, $P(n)$ becomes supercritical, and the maximal connected circuit size becomes linear in n (Bollobás, 1984), which makes it linguistically implausible for large n.

For $c = 0.1$, the polarity probability is $p(n) = 0.1/n$. The inter-neuron storage connectivity probability is $P(n)$, as specified

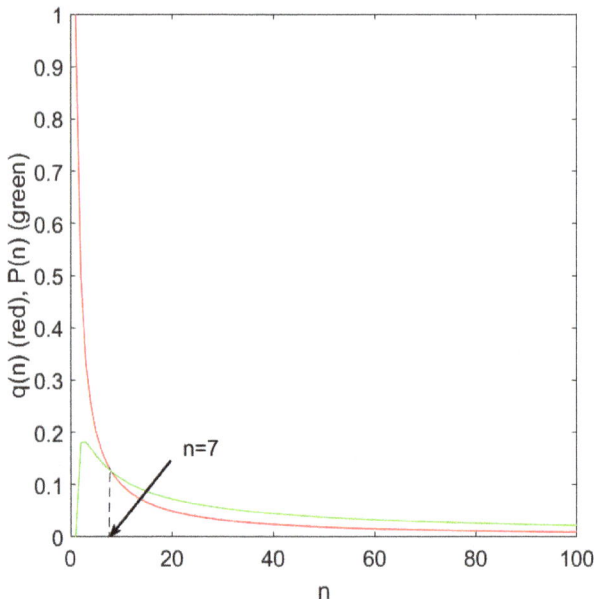

Figure 3.8.   For $p(n) = c/n$ with c=0.5, the inter-neuron storage connectivity probability is $P(n)$ (green curve) and the critical connectivity probablilty is $q(n)$ (red curve). The critical point satisfying $P(n) = q(n) = 1/n$ is at $n = 7$ (hence, $[\log(n)] = 2, C(n) = [n/\log(n)] = 3$, Baram, 2020b).

by Eq. (3.24). The critical point satisfying $P(n) = q(n) = 1/n$ is found to be $n = 22,030$, as depicted in Fig. 3.9. The corresponding maximal connected circuit size is $s = [\log(22,030)] = 10$, which is the critical linguistic limit of the neural storage polarity code word size. For $n > 22,030$, $P(n)$ becomes supercritical and the maximal connected circuit size becomes linear in n (Bollobás, 1984), which makes it linguistically implausible.

It follows that while, for large neural assemblies under supercritical polarity probability, the connected neural circuits are, rather uniformly, large, producing linguistically implausible long codewords, the range of subcritical storage polarity probabilities produces a variety of linguistically plausible, relatively short (2–10 letters) code words. The smaller the value of c (hence, of $p(n)$), the greater the linguistically plausible range.

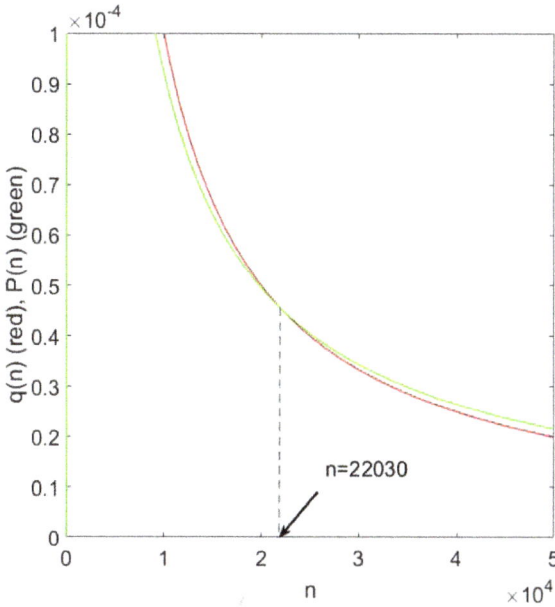

Figure 3.9. For $p(n) = c/n$ with c=0.1, the critical point satisfying $P(n) = q(n) = 1/n$ is at $n = 22,030$ (hence, $[\log(n)] = 10$, $[n/\log(n)] = 2,203$, Baram, 2020b)).

Given an assembly of n neurons under subcritical storage polarity probability, the minimal subcritical Hebbian storage capacity, represented by Eq. (3.21), $B(n) = n$, is greater than the polarity controlled capacity represented by Eq. (3.5), $C(n) = n/\log(n)$. However, in contrast to the polarity controlled case, which, for appropriate value of c $(0 < c < 1)$, is subcritical for any value of n, Hebbian storage subcriticality has a limited range depending on the value of c, as illustrated by Figs. 3.8 and 3.9.

3.6 Discussion

Linguistic considerations, such as the number and the sizes of code words, appear to be as significant for the language of cortical information representation as they are for natural and artificial languages. Extending fundamental results from the theory of random graphs, we have shown that subcritical storage probability of positive polarity holds the key to small circuit segregation, which, in turn, holds the key to linguistically plausible short code words. In contrast, we have shown that supercritical probability of positive polarity, yielding inherently long code words, misrepresents efficient cortical linguistics. The subcritical linguistic range of cortical codewords (hence, neural circuits), resembling the one found in natural languages is between 2 and 10 "letters", represented by neuronal membrane polarity states. Finally, it might be noted that Hebbian storage in a circuit of 10 neurons is in close proximity to "the magical number 7 plus or minus 2", which has been suggested as an upper limit for working memory capacity (Miller, 1956; Pribram *et al.*, 1960).

Chapter 4

Hebbian Random Graphs and Firing-rate Dynamics

4.1 Introduction

Random graphs normally assume a binary connectivity between nodes (Erdos and Renyi, 1959, 1960; Gilbert, 1959). While such connectivity has been adopted by mathematical neural circuit models (McCulloch and Pitts, 1943; Amari, 1972; Hopfield, 1982), it is recognized as molecularly induced membrane (Melnick, 1994) and synapse (Atwood and Wojtowicz, 1999) potential values being above or below a certain resting potential (about −60 mV). As shown in this chapter, the widely accepted Hebbian paradigm of learning and memory (Hebb, 1949), while mathematically formalized in binary terms (McCulloch and Pitts, 1943; Amari, 1972; Hopfield, 1982), adds another dimension to the corresponding graph definition, namely, the number of polarity patterns stored. This, as we show, has a fundamental effect not only on the corresponding synaptic weights, but also on cortical linguistics characterized by the length and number of the corresponding code words. Small segregated circuits are addressed as examples, since larger circuits would present unmanageable requirements on Hebbian matrix calculations and graphical circuit representation. Regardless, interest in small neural circuit ("small-world") as a possible cortical reality is gaining considerable interest (Bassett *et al.*, 2006). Following the findings presented in Chapter 3, Baram (2020b) has further demonstrated the combined effects of Hebbian storage on two different forms of cortical linguistics, namely, circuit connectivity and firing-rate dynamics, as discussed and illustrated in this chapter.

4.2 Random Hebbian criticality effects on circuit linguistics

Consider a circuit of n neurons, whose membrane polarity states, $w_j^{(i)} \in (0,1)$, $j = 1, \ldots, n$ at discrete times, $i = 1, \ldots, M$, take the value 1 with probability p. Storage of polarity state values is performed according to the so-called Hebbian rule (e.g. Hopfield, 1982)

$$\mathbf{w}_{k,j} = \sum_i^M w_k^{(i)} w_j^{(i)} \tag{4.1}$$

Retrieval of polarity vectors is performed by the rule (McCulloch and Pitts, 1943)

$$x_k = \begin{cases} 0, & \text{if } u_k \leq 0 \\ 1, & \text{if } u_k > 0 \end{cases} \tag{4.2}$$

where

$$u_k = \sum_{j=1}^n w_{k,j} z_j \tag{4.3}$$

is the input to the k'th neuron, where z_j, $j = 1, \ldots, n$, are the components of a similarly binary probe.

Early results on the memory and error correction capacities of binary neural networks have linked high capacities to large networks (McEliece *et al.*, 1987; Kuh and Dickinson, 1989; Baram and Sal'ee, 1992), assumed to be fully connected by the underlying model (Hopfield, 1982). However, such large networks seem to contradict the notion of segregated cortical information and function, supported by numerous experimental studies.

We next demonstrate that neural circuit segregation is a natural outcome of the Hebbian storage of randomly generated polarity patterns. Let $n = 10, p = 0.5/n$ and $M = 10$. It can be seen

(a)

$$\mathbf{W} = \begin{bmatrix} 1 & 0 & 0 & 0 & 0 & 0 & 0 & 0 & 0 & 0 \\ 0 & 1 & 0 & 0 & 0 & 0 & 0 & 0 & 1 & 0 \\ 0 & 0 & 0 & 0 & 0 & 0 & 0 & 0 & 0 & 0 \\ 0 & 0 & 0 & 1 & 0 & 0 & 0 & 0 & 0 & 0 \\ 0 & 0 & 0 & 0 & 0 & 0 & 0 & 0 & 0 & 0 \\ 0 & 0 & 0 & 0 & 0 & 1 & 0 & 0 & 0 & 0 \\ 0 & 0 & 0 & 0 & 0 & 0 & 0 & 0 & 0 & 0 \\ 0 & 0 & 0 & 0 & 0 & 0 & 0 & 1 & 0 & 0 \\ 0 & 1 & 0 & 0 & 0 & 0 & 0 & 0 & 1 & 0 \\ 0 & 0 & 0 & 0 & 0 & 0 & 0 & 0 & 0 & 0 \end{bmatrix}$$

(b)

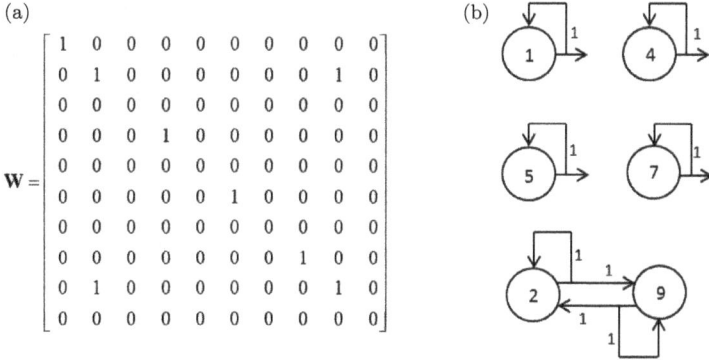

Figure 4.1. Subcritical Hebbian segregation corresponding to $n = 10$, $p = 0.5/n$ and $M = 10$. (a) Synaptic weights matrix and (b) segregated neural circuits (Baram, 2020b).

that, in terms of random graph theory (Erdos and Renyi, 1959, 1960), p is subcritical. The storage of M n-dimensional randomly generated polarity vectors according to Eq. (4.1) has produced the synaptic weights matrix $\mathbf{W}$ depicted in Fig. 4.1(a). The connectivity represented by this matrix can be seen to translate into five segregated (externally disconnected) circuits depicted in Fig. 4.1(b), whose neurons are numbered according to the rows (or columns) of the matrix $\mathbf{W}$. Four subcircuits consist of one neuron each (neurons 1, 4, 5 and 7), while one subcircuit consists of two neurons (neurons 2 and 9). As implied by Fig. 4.1(a), all synaptic weights are valued at 1.

Now let $n = 10$, $p = 1.5/n$ and $M = 10$, making p supercritical by graph theoretic terminology. The storage of M n-dimensional randomly generated polarity vectors according to Eq. (4.1) has produced the synaptic weights matrix $\mathbf{W}$ depicted in Fig. 4.2(a). With some of the weights exceeding the value 1, the connectivity represented by this matrix can be seen to translate into three segregated (externally disconnected) circuits depicted in Fig. 4.2(b), whose neurons are numbered according to the rows (or columns) of the matrix $\mathbf{W}$. One of these subcircuits consists of a single neuron (neuron 4), one consists of three neurons (neurons 2, 9 and 10) and one consists of four neurons (neurons 3, 5, 7 and 8).

It should be noted that, under subcriticality (hence, highly sparse polarity patterns), for $M = n(=10)$ all stored polarity vectors are mutually orthogonal, which allows for equilibrium capacity M, maintaining the maximal connected subcircuit size implied be Eq. (2.5).

The transition from subcritical p ($p = 0.5/n$) to supercritical p ($p = 1.5/n$) has resulted, then, in a dramatic linguistic change. In the subcritical case, the maximal segregated circuit size of 2 satisfies the condition represented by Eq. (2.5) (as $2 < \log(10) = 2.3$), while in the supercritical case maximal segregated circuit size of 4 does not satisfy the condition (as $4 > \log(10)$). A more general analysis will be presented in the following chapter.

Next, let $n = 10$ and $p = 0.5/n$ as before, but let M increase to 20. The storage of 20 n-dimensional randomly generated polarity vectors according to Eq. (4.1) has produced the synaptic weights matrix $\mathbf{W}$ depicted in Fig. 4.3(a), which can be seen to have elements greater than 1. The connectivity represented by this matrix translates into

(a)

$$
\mathbf{W} = \begin{bmatrix}
0 & 0 & 0 & 0 & 0 & 0 & 0 & 0 & 0 & 0 \\
0 & 2 & 0 & 0 & 0 & 0 & 0 & 0 & 0 & 1 \\
0 & 0 & 1 & 0 & 0 & 0 & 1 & 1 & 0 & 0 \\
0 & 0 & 0 & 1 & 0 & 0 & 0 & 0 & 0 & 0 \\
0 & 0 & 0 & 0 & 2 & 0 & 1 & 0 & 0 & 0 \\
0 & 0 & 0 & 0 & 0 & 0 & 0 & 0 & 0 & 0 \\
0 & 0 & 1 & 0 & 1 & 0 & 2 & 1 & 0 & 0 \\
0 & 0 & 1 & 0 & 0 & 0 & 1 & 2 & 0 & 0 \\
0 & 0 & 0 & 0 & 0 & 0 & 0 & 0 & 2 & 1 \\
0 & 1 & 0 & 0 & 0 & 0 & 0 & 0 & 1 & 2
\end{bmatrix}
$$

(b)

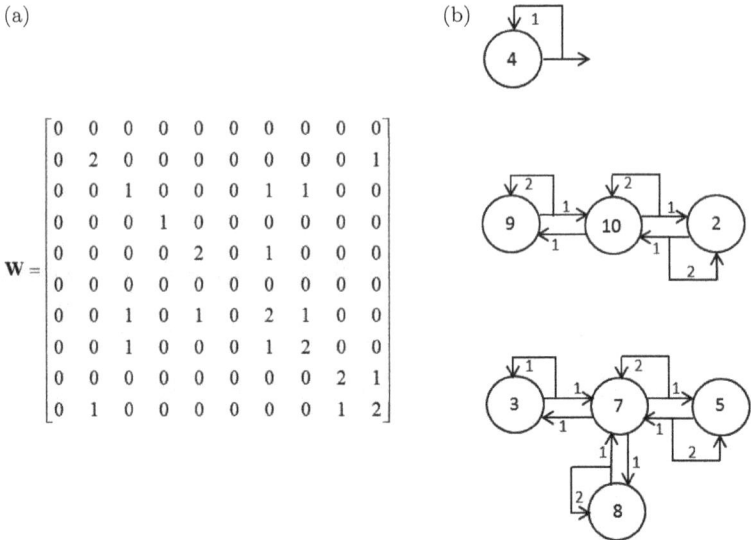

Figure 4.2. Supercritical Hebbian segregation corresponding to $n = 10$, $p = 1.5/n$ and $M = 10$. (a) Synaptic weights matrix and (b) segregated neural circuits (Baram, 2020b).

the four segregated circuits depicted in Fig. 4.3(b), whose neurons are numbered according to the rows (or columns) of the matrix $\mathbf{W}$. It can be seen that three of the circuits consist of a single neuron each (neurons 1, 6 and 8) and one circuit consists of three neurons (neurons 2, 4 and 9). While p is subcritical from a graph theoretic viewpoint, as in the case depicted by Fig. 4.1 ($p = 0.5/n$), the maximal segregated circuit size of 3 violates the condition Eq. (2.5) (as $3 > \log(10) = 2.3$). The reason is that, in contrast to the case $M = n(=10)$ corresponding to Fig. 4.1, the present case of $M = 20 > n = 10$ does not yield full mutual orthogonality nor capacity of M stored polarity patterns, resulting in excessively large connectivity in spite of the subcritical value of p.

4.3 Firing-mode reproduction by polarity recall

Neuronal firing-rate models (Lapicque, 1907; Hodgkin and Huxley, 1952; Gerstner, 1995) have been put in discrete time forms for computational purposes (Baram, 2017a,b, 2018). The nature of neural circuit firing-rate dynamics, having direct impact on cortical function, is closely associated with synaptic plasticity (Bienenstock *et al.*, 1982). The latter, in contrast to circuit polarity, is not binary. The convergence of time-varying synaptic plasticity weights

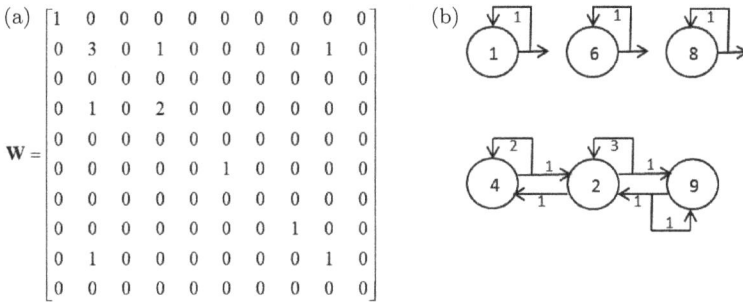

Figure 4.3. Hebbian segregation under increased Hebbian storage, corresponding to $n = 10$, $p = 0.5/n$ and $M = 20$. (a) Synaptic weights matrix and (b) segregated neural circuits. While p is subcritical, the maximal connected subcircuit size exceed the limit of Eq. (2.5) due to excessive Hebbian storage (Baram, 2020b).

to limit values results in a high variety of dynamic firing-rate modes, characterized by a code of global attractors (Baram, 2013a). However, as we argue and demonstrate next, given neuronal internal properties, a neural circuit's firing-rate dynamics are completely determined by a vector-valued probe, retrieving a memorized polarity pattern by the mechanism represented by Eqs. (4.1)–(4.3). This implies that once a neural circuit is probed, revealing a stored polarity vector, it can only evolve into one final state of synaptic plasticity weights. Consequently the circuit's firing-rate dynamics can only evolve into one final set of dynamic modes, which may be synchronous and identical for all the circuit's neurons, or unequally and asynchronously produced by different neurons.

The discrete-time firing-rate model for the individual neuron is (e.g., Baram, 2013a)

$$v_i(k) = \alpha_i v_i(k-1) + \beta_i f(\boldsymbol{\omega}_i^T(k)\boldsymbol{v}_i(k-1) + u_i) \qquad (4.4)$$

where $v_i(k)$, is the neuron's firing-rate, $\boldsymbol{v}_i(k)$ is the vector of firing rates of the neuron's pre-neurons, $\alpha_i = \exp(-1/\tau_{m_i})$ and $\beta_i = 1 - \alpha_i$, with τ_{m_i} the membrane time constant of the neuron, $\boldsymbol{\omega}_i(k)$ is the vector of synaptic weights corresponding to the neuron's pre-neurons (including self-feedback), $u_i = I_i - r_i$, with I_i the neuron's circuit-external activation input and r_i the neuron's membrane resting potential and f_i the conductance-based rectification kernel defined by Carandini and Ferster (2000)

$$f_i(x) = f(x) = \begin{cases} x & \text{if } x \geq 0 \\ 0 & \text{if } x < 0 \end{cases} \qquad (4.5)$$

The BCM plasticity rule (Bienenstock *et al.*, 1982) now takes the multi-neuron form (Baram, 2017a,b, 2018)

$$\boldsymbol{\omega}_i(k) = \varepsilon_i \boldsymbol{\omega}_i(k-1) + \gamma_i[v_i(k-1) - \theta_i(k-1)]\boldsymbol{v}_i^2(k-1) \qquad (4.6)$$

where $\boldsymbol{v}_i^2(k-1)$ is the vector whose components are the squares of the components of $\boldsymbol{v}_i(k-1)$, and for sufficiently large N to guarantee convergence of the sum,

$$\theta_i(k) = \delta_i \sum_{\ell=0}^{N} \exp(-\ell/\tau_{\theta_i})v_i^2(k - \ell) \tag{4.7}$$

with $\varepsilon_i = \exp(-1/\tau_{w_i})$, $\gamma_i = 1 - \varepsilon_i$, $\delta_i = 1/\tau_{\theta_i}$ where τ_{w_i} and τ_{θ_i} are time constants.

Next, we examine and demonstrate the effects of synapse polarization on circuit structure and firing dynamics. The essence of these findings will be demonstrated for small circuits of two neurons, as the simulation of large circuit firing is highly elaborate, tedious and space consuming. Consider a 2-neuron circuit having the following identical parameters

$$u_1 = u_2 = 1, \ \tau_{m,1} = \tau_{m,2} = 1, \ \tau_{w,1} = \tau_{w,2} = 5, \ \tau_{\theta,1} = \tau_{\theta,2} = 1.$$

Starting with full connectivity, changes in circuit connectivity due to synapse silencing are illustrated in Fig. 4.4. Different initial values of the firing-rates, $v_1(0)$ and $v_2(0)$ were applied to check robustness of the results and the claims made. The resulting neuronal firing modes, simulated by running Eqs. (4.4)–(4.7), are displayed in Fig. 4.5.

We establish the Hebbian memory of the polarity patterns in the three cases depicted in Fig. 4.4 by employing the mechanism

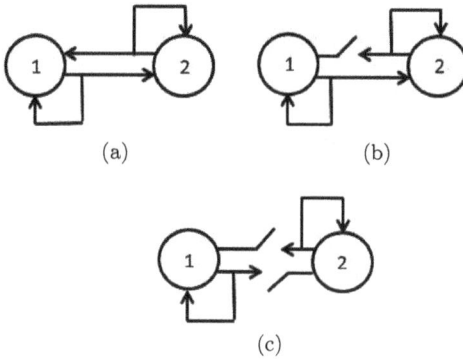

(a)   (b)

(c)

Figure 4.4. Two-neuron circuit modification and segregation by synapse silencing. (a) Fully connected circuit, (b) asynchronous circuit connectivity and (c) circuit segregation into two single isolated neurons by synapse silencing (Baram, 2020b).

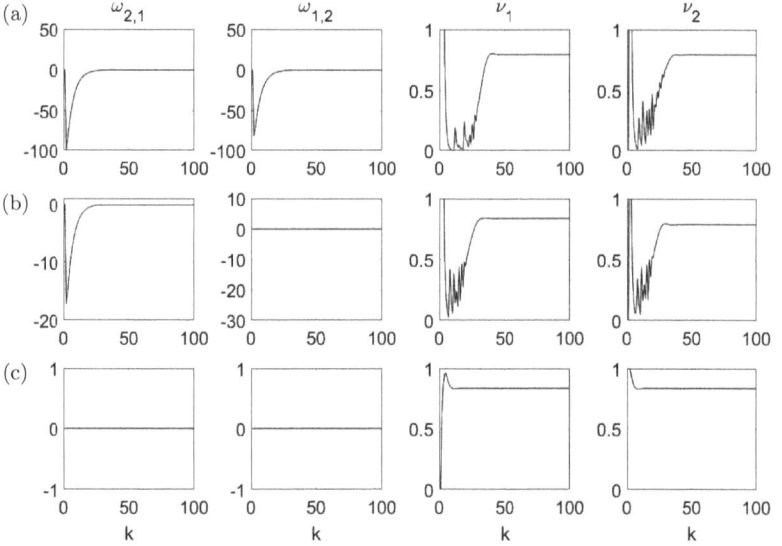

Figure 4.5. Temporal plasticity evolution $\omega_{2,1}$ and $\omega_{1,2}$ and firing-rate sequences, v_1 and v_2, corresponding to the three circuits depicted in Fig. 4.4. The plasticity synaptic weights were initialized at the convergence values of the polarity synaptic weights (Baram, 2020b).

represented by Eqs. (4.1)–(4.3). We assume for illustrative purposes that in each of the cases we have a synaptic weights matrix which, calculated by Eq. (4.1) in two time steps, corresponds to the circuit polarity states displayed in Fig. 4.4.

In case (a) we have

$$\mathbf{w} = 2\begin{pmatrix} 1 & 1 \\ 1 & 1 \end{pmatrix} = \begin{pmatrix} 2 & 2 \\ 2 & 2 \end{pmatrix} \tag{4.8}$$

This is the initial condition in the calculation of the time-varying plasticity weights according to Eq. (4.6), and reaching the limit

$$\omega_{1,1} = \omega_{1,2} = \omega_{2,1} = \omega_{2,1} = -0.1291.$$

In case (b) we have

$$\mathbf{w} = 2\begin{pmatrix} 1 & 0 \\ 1 & 1 \end{pmatrix} = \begin{pmatrix} 2 & 0 \\ 2 & 2 \end{pmatrix} \tag{4.9}$$

which, in the appropriate positions, represents the initial condition for the continuous-time plasticity weights. The limit values of the plasticity weights calculated according to Eq. (4.6) were found to be

$$\omega_{1,1} = -0.1216, \ \omega_{1,2} = -0.1375, \ \omega_{2,1} = 0, \ \omega_{2,2} = -0.1925$$

In case (c) we have

$$\mathbf{w} = 2 \begin{pmatrix} 1 & 0 \\ 0 & 1 \end{pmatrix} = \begin{pmatrix} 2 & 0 \\ 0 & 2 \end{pmatrix} \tag{4.10}$$

which constitutes the initial condition for the continuous plasticity weights. The final values of the plasticity weights calculated according to Eq. (4.6) were found to be $\omega_{1,1} = -0.1925, \ \omega_{1,2} = 0, \ \omega_{2,1} = 0,$ $\omega_{2,2} = -0.1925$.

The values of the synaptic weight $\omega_{2,1}, \omega_{1,2}$ and the firing-rate v_1, v_2 sequences in the three cases are displayed in Fig. 4.5. It can be seen that, subject to initial transients due to different initial conditions, neurons in circuits that have symmetric connectivity (cases (a) and (c)) converge to the same firing-rate dynamics, while the neurons in a circuit which has a non-symmetric connectivity (case (b)) converge to different firing-rate dynamics. Indeed, the final values of v_1 and v_2 in the three cases are:

(a) $v_1(k = 100) = v_2(k = 100) = 0.7947$ (symmetric, as original polarity)
(b) $v_1(k = 100) = 0.7888, v_2(k = 100) = 0.8385$ (asymmetric, as original polarity)
(c) $v_1(k = 100) = v_2(k = 100) = 0.8385$ (symmetric, as original polarity)

Subject to internal neural parameters, the cortical languages of polarity, on the one hand, and firing-rate dynamics, on the other, are, then, interchangeable, representing such neural circuit properties as symmetry, synchrony and asynchrony. The expression of circuit polarity and such dynamic properties in a variety of firing-rate modes has been noted (Baram, 2013a).

4.4   Discussion

This chapter has essentially suggested two new paradigms concerning information representation, memory and dynamic manifestation by neural circuits. The first paradigm is that of random circuit segregation into smaller circuits. Natural and artificial (e.g. computer) languages, facilitating an incredible wealth of expression, employ short words. In cortical contexts, while letters, or bits, may be represented by polarity states, short words would be represented by relatively small neural circuits. As we have shown, the segregation of many neurons into small circuits can be done by Hebbian storage of random polarity patterns. The linguistic use of such segregation, characterized by the length and number of the resulting polarity code words, requires an extension of graph theory to what we call *Hebbian graphs*, constructed by the Hebbian learning paradigm. Random circuit segregation by the Hebbian learning rule complements forced segregation suggested and analyzed in our previous work (Baram, 2017a,b, 2018), controlling circuit connectivity and firing-rate dynamics. Such segregation will also be considered in subsequent sections of this book.

The second new paradigm suggested in this chapter is that a meaningful representation of cortical information, having a fundamental impact on memory and other cortical functions, is divided into two domains: the polarity domain, which, dominating circuit connectivity, consists of binary representation, and the continuous plasticity domain, dominating firing-rate dynamics. The fascinating aspect of these seemingly separate domains is that they actually combine at a very specific point of time and action. As the binary information of polarity is stored by so-called Hebbian learning, forming memory, it constitutes the starting point of continuous plasticity, which, by convergence of synaptic weights, defines the dynamic attractor of firing-rate, instrumental in the execution of cortical function. The correspondence between neural circuit polarity, firing-rate attractors and cortical functions has been noted (Baram, 2017a, b). However, the possible role of Hebbian polarity memory in the determination of firing-rate dynamics appears to be suggested here for the first time. Finally, it might be noted that while publications of experimental results on membrane and synapse polarities must report the specific cortical region addressed by a specific experiment, the notion of polarity, referring to the corresponding potential being above or below a certain value, would seem to apply more generally to any cortical region.

Chapter 5

Synaptic Polarity and Primal-size Categories of Neural Circuit Codes

5.1 Introduction

The state of neuronal activity, contrasted by neuronal silence, has been found to depend on the somatic membrane potential being above or below a certain threshold value (about $-60\,\mathrm{mV}$, Melnick 1994). A functional role for the silent neuron has been noted in the 2014 Nobel prize-winning work on cortical maps of place (O'Keefe and Dostrovsky 1971; Hafting *et al.*, 2005). The state of synaptic transmissivity, contrasted by synaptic silence, has been found to depend on the value of pre-synaptic membrane potential, controlled by external stimulation and molecular properties, with respect to a certain threshold value (also about $-60\,\mathrm{mV}$, Atwood and Wojtowicz, 1999). A detailed biophysical model relates long-term synaptic potentiation and long-term synaptic depression, which are also viewed in a binary ("bidirectional") context, to the variable properties and relative numbers of AMPA and NMDA receptors and their external stimulation (Castellani *et al.*, 2001). The effects of neuronal circuit polarity on cortical connectivity, firing dynamics and memory have been analyzed (Baram, 2018).

Following Baram (2020a), we are motivated by a desire to understand the correspondence between neural circuit connectivity structure and the categorization of the information it might convey. The underlying structure is shown to be governed by two experimentally supported pairs of paradigms. The first is the above mentioned pair of polarity paradigms, namely, membrane polarity and synapse polarity. The second is the pair of neuronal self-feedback paradigms, namely, the axonal discharge self-feedback paradigm (Carlsson and Lindquist, 1963; Smith and Jahr, 2002) and the synaptically-mediated self-feedback paradigm (Groves *et al.*, 1975). While, from a mathematical viewpoint, the second paradigm is shown to contain the first, their different experimental origins have been associated with different molecular and neurophysiological mechanisms. Formulating the polarities of neuronal membranes and synapses as on/off gates, the polarity code of a fully connected circuit is defined as the set of all neuronal polarity permutations. Employing a mathematical concept used for grouping graphical objects sharing certain attributes (Awodey, 2010), we call a neural circuit polarity code, or subcode, sharing a certain substructure, a *category*. In particular, we investigate categories of primal-size which, being indivisible, endows the category members with a common characterization. When the circuit polarity code size is not a prime number, the code factorizes into primal-size categories. Circuits of small sizes can interact in a feedforward fashion, arousing relatively large cortical areas. The effects of different neural circuit polarity categories on the neuronal firing-rate modes are illustrated by example. The analytically derived neural circuit polarity codes and category sizes are found to yield, explain and extend information capacity values experimentally observed in behavioral studies, such as "the magical number seven" (Miller, 1956) and "the magical number four" (or, more specifically, "between three and five", Cowan, 2001), associated with so-called "working memory" (Pribram *et al.*, 1960). The dimensionality associated with such capacities is shown to result from prime factorization of composite circuit polarity code sizes. While these have been previously argued on grounds of psychological experiments, here they are further supported on analytic grounds by the so-called Hebbian memory paradigm.

5.2   Polarity codes and primal-size categories under axonal discharge self-feedback

Neuronal membrane and synapse polarities will be described as on/off gates which represent, in the "off" ("disconnect") position, negative polarity, and, in the "on" ("connect") position, positive polarity (Baram, 2018). Axonal discharge neuronal self-feedback has been observed and its molecular implementation has been specified (Carlsson and Lindquist, 1963; Smith and Jahr, 2002). The axonal discharge feeds back directly as an input to the neuron, without being further gated by a synapse.

Graphically, the axonal discharge polarity, which is identical to that of the somatic membrane, will be represented by a direct line connecting the neuron's axon to the neuron's input. The axonal discharge self-feedback neuron has two polarity states, graphically represented by Figs. 5.1(a) and 5.1(b). When the membrane is active (Fig. 5.1(a)), so is the axonal discharge self-feedback, and when the membrane is silent (Fig. 5.1(b)), there is no self-feedback. As the two states, (a) and (b), are the most elementary polarity patterns under axonal discharge self-feedback, they may be termed the "letters" of the neural circuit polarity code in this context. Yet, as the isolated neuron is also a "1-neuron circuit", the two polarity states also constitute, in this specific case, "1-letter words" of the corresponding polarity code. As 2 is a prime number, the two circuits (a) and (b) in

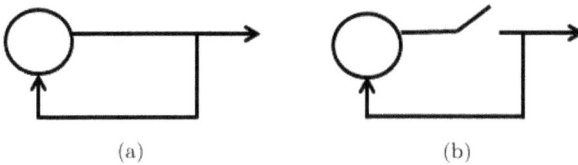

(a)   (b)

Figure 5.1.   1-category polarity code of primal-size 2 of a 1-neuron circuit with axonal discharge self-feedback. While the membrane polarity state is "on" in circuit (a) and "off" in circuit (b), the self-feedback in both circuits is permanently in the "on" state (Baram, 2020a).

Fig. 5.1 constitute a circuit polarity category. Primality guarantees that a polarity category of n-neuron circuits cannot be divided into smaller subsets having the same number of circuits, unless this number is 1.

For instance, the 2-circuit set depicted in Fig. 5.1 is of primal size, as, while it divides into two subsets, each of these subsets consists of a single circuit. The 1-category polarity code of primal size 7 of a 2-neuron circuit with axonal discharge self-feedback is depicted in Fig. 5.2.

In order to derive the n-neuron circuit polarity code under axonal discharge self-feedback, it is first noted that there can be up to n active neurons (neurons with positive membrane polarities) and $n(n-1)$ synapses. A synapse can be either active or silent. It follows that there can be up to

$$P^{(1)}(n) = 2^{n(n-1)} \tag{5.1}$$

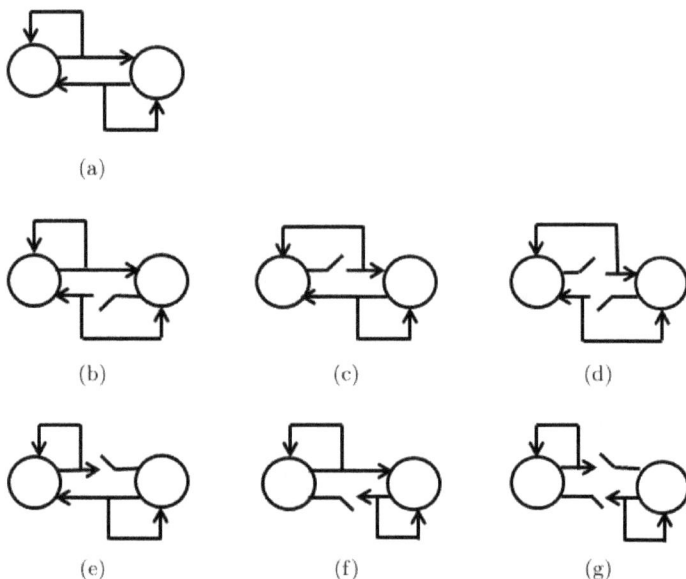

(a)

(b)   (c)   (d)

(e)   (f)   (g)

Figure 5.2. The primal-size 7-word polarity code constituting a single category of 2-neuron circuits under axonal discharge self-feedback (Baram, 2020a).

different circuits involving only active membranes and possibly silent synapses. In addition, there can be up to

$$P^{(2)}(n) = \binom{n}{1} + \binom{n}{2} + \cdots + \binom{n}{n} = 2^n - 1 \qquad (5.2)$$

circuits with silent neurons (counting all permutations of 1 silent neuron, 2 silent neurons, ..., n silent neurons out of n neurons). Note that the circuits corresponding to the active states of these neurons were already counted in $P^{(1)}(n)$, so that the only additional circuits are the same ones with these neurons silenced. It follows that there is a maximum total of

$$P(n) = P^{(1)}(n) + P^{(2)}(n) = 2^{n(n-1)} + 2^n - 1 \qquad (5.3)$$

different circuits, constituting the size of the circuit polarity code of n axonal discharge self-feedback neurons.

Equation (5.3) yields $P(1) = 2$, $P(2) = 7$, $P(3) = 71$ and $P(4) = 4111$. The neural circuit polarity code of two neurons with axonal discharge self-feedback is depicted in Fig. 5.2. As 7 is a prime number, this polarity code constitutes a category. Employing methods of primality verification, e.g., the Miller-Rabin method (Miller 1976; Rabin 1980), it can be verified that $P(1), P(2), P(3)$ and $P(4)$ are all prime numbers (they are the 1st, the 4th, the 20th and the 566th prime numbers, not counting 1). On the other hand, $P(5) = 1,048,607$, $P(6) = 1,073,741,887$ and $P(7) = 4,398,046,511,231$ are not prime. As the prime factorization of 1,048,607 is $7 \times 59 \times 2,539$, the polarity code of 5-neuron circuits factorizes into $7 \times 59 = 413$ categories of primal size 2,539, or $7 \times 2,539 = 17,773$ categories of primal size 59, or $59 \times 2,539 = 149,801$ categories of primal size 7. As the prime factorization of 1,073,741,887 is $37 \times 109 \times 266,239$, the polarity code of 6-neuron circuits factorizes into $37 \times 109 = 4,251$ categories of primal size 266,239, or $37 \times 266,239 = 9,850,843$ categories of primal size 109, or $109 \times 266,239 = 29,020,051$ categories of primal size 37. As the prime factorization of 4,398,046,511,231 is $163 \times 15,451 \times 1,746,287$, the polarity code of 7-neuron circuits factorizes into $163 \times 15,451 = 2,518,513$ categories of primal size 1,746,287, or $163 \times 1,746,287 = 284,644,781$ categories of primal

size 15,451, or 15,451 × 1,746,287 = 26,981,880,437 categories of
primal size 163. Our attempts to derive the prime factorization of the
polarity code sizes corresponding to n-neuron circuits for $n \geq 8$ have
failed, as the term $2^{n(n-1)}$ in Eq. (5.3) becomes exceedingly dominant
with respect to the other terms, erroneously producing powers of
2 (specifically, $2^{n(n-1)}$) as "prime factors" of $P(n)$. In contrast to
the single primal polarity code sizes corresponding to circuits of 1–4
neurons, which may be characterized as having a single dimension,
the three different primal-size polarity category sizes corresponding
to circuits of 5, 6 and 7 neurons may be characterized as representing
three different dimensions.

Composite numbers that factorize into three prime numbers have
been called sphenic numbers. Such numbers have been addressed
in purely mathematical contexts (Lehmer, 1936) and their natural
relevance appears to be noted here for the first time.

The entire code of an n-neuron circuit polarity under axonal
discharge self-feedback can be cortically represented by $P(n)$ circuits.
For instance, the 7-circuit group of 2-neuroncircuits depicted in
Fig. 5.2 represents the corresponding seven different polarity code
words under axonal self-feedback. The implementation of the polarity
code of a 1-neuron circuit would require then $1 \times 2 = 2$ neurons,
while the implementation of the polarity code of a 2-neuron circuit
would require $2 \times 7 = 14$ neurons. The implementation of the
polarity codes of 3 and 4-neuron circuits would require $3 \times 71 = 213$
and $4 \times 4{,}111 = 16{,}444$ neurons, respectively. The implementation
of the polarity codes of 5 and 6-neuron circuits would require
$5 \times 1{,}048{,}607 = 5{,}243{,}035$ and $6 \times 1{,}073{,}741{,}887 = 6{,}442{,}451{,}322$
neurons, respectively. The implementation of a 7-neuron circuit
polarity code, having size $P(7) = 4{,}398{,}046{,}511{,}231$ would require
$7 \times 4{,}398{,}046{,}511{,}231 = 30{,}786{,}325{,}578{,}617$ neurons, far exceeding
the number of neurons in the human brain (10^{12}). However, it should
be noted that any fraction of a polarity code, including such codes
corresponding to circuits of n neurons where $n \geq 7$, is implementable
by a number of categories smaller than the one calculated for the full
implementation of the polarity code. For instance, if only half the
number of neurons needed for a full implementation of the 7-neuron

Table 5.1. Neural circuit polarity code sizes for circuit sizes 1–7 under axonal discharge self-feedback. The polarity category sizes are of primal size, 2, 7, 71 and 4,111, respectively, for circuits of 1–4 neurons, and sphenic (3-prime factorizable) for circuits of 5, 6 and 7 neurons (Baram, 2020a).

# of circuit neurons	Polarity code size	Polarity code size prime factorization	# of primal-size categories	Primal category size
1	2	2	1	2
2	7	7	1	7
3	71	71	1	71
4	4111	4111	1	4111
5	1,048,607	$7 \times 59 \times 2539$	413	2539
			17,773	59
			149,801	7
6	1,073,741,887	$37 \times 109 \times 266,239$	4251	266,239
			9,850,843	109
			29,020,051	37
7	4,398,046,511,231	$163 \times 1545 \times 1,746,287$	2,518,513	1,746,287
			284,644,781	15451
			26,981,880,437	163

polarity code are available, half the number of categories in each of the three corresponding sphenic subgroups of categories would be implementable.

Table 5.1 summarizes the circuit sizes, the corresponding circuit polarity code sizes, the polarity code size prime factorization, the numbers of primal-size categories, and the corresponding primal category sizes for circuits of 1–7 neurons. The polarity code sizes of 5, 6 and 7-neuron circuits are factorized as specified above.

5.3 Polarity codes and primal-size categories under synaptic self-feedback

Neuronal self-feedback, employing an intermediate synapse between axonal output and neuronal input, has been suggested (Groves *et al.*, 1975) as an alternative to direct axonal discharge self-feedback without synaptic mediation (Carlsson and Lindquist, 1963; Smith and Jahr, 2002). In contrast to axonal discharge self-feedback, under the synaptic self-feedback paradigm, each of the circuit neurons has a self-synapse that can be in one of the two polarity states: "on" or "off". It follows that there are

$$C(n) = \binom{n}{0} + \binom{n}{1} + \cdots + \binom{n}{n} = 2^n \qquad (5.4)$$

permutations of self-feedback polarity states in a circuit of n neurons. The total polarity code size for an n-neuron circuit is then

$$Q(n) = C(n)P(n) \qquad (5.5)$$

where $P(n)$ is the polarity code size of n-neuron circuits under the axonal discharge feedback paradigm, represented by Eq. (5.3). Specifically, for $n = 1$ we have $C(1) = 2$, hence, $Q(2) = 2 \times 2 = 4$. For $n = 2$ we have $C(2) = 4$, hence, $Q(2) = 4 \times 7 = 28$. For $n = 3$ we have $C(3) = 8$, hence, $Q(3) = 8 \times 71 = 568$, and for $n = 4$ we have $C(4) = 16$, hence, $Q(4) = 16 \times 4111 = 65{,}776$. For $n = 5$ we have $C(5) = 32$, hence, $Q(5) = 32 \times 1{,}048{,}607 = 33{,}555{,}424$, and for $n = 6$ we have $C(6) = 64$, hence, $Q(6) = 64 \times 1{,}073{,}741{,}887 = 68{,}719{,}480{,}768$. For $n = 7$ we have $C(7) = 128$, hence, $Q(7) = 128 \times 4{,}398{,}046{,}511{,}231 = 562{,}949{,}953{,}437{,}568$. Similarly, for each of the primal-size categories of n-neuron circuits with $n = 5$, $n = 6$ and $n = 7$, produced by prime factorization under the axonal discharge self-feedback paradigm, there are $C(n)$ such categories associated with the permutations of the polarity states corresponding to synaptic self-feedback.

The 4-circuit polarity code for a single neuron under the synaptic self-feedback paradigm is displayed in Fig. 5.3, where the two categories of primal size 2 each are represented by columns (a) and (b). The 4-circuit polarity categories of 7 circuits of 2 neurons each are displayed in Fig. 5.4. It can be seen that the first category, represented by column (a), is the category corresponding to axonal discharge self-feedback, hence both neurons' self-synapses being in the "on"" state, as derived in Section 5.2. The second category, represented by column (b), corresponds to the left neuron's self-synapse being in the "off" state, while the right neuron's self-feedback is in the "on" state. The third category, represented by column (c), corresponds to the right neuron's self-synapse being in the "off" state,

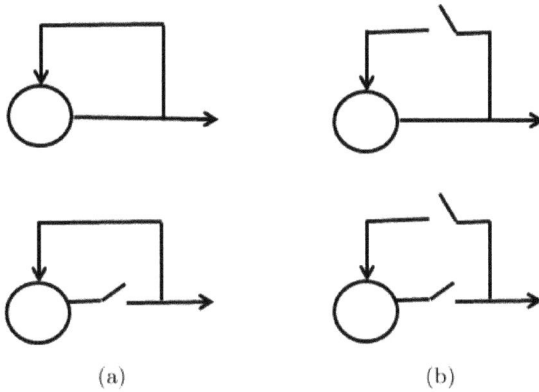

(a)   (b)

Figure 5.3. The 2-circuit categories, represented by columns (a) and (b), of 1-neuron polarity code under synaptic self-feedback. Category (a) is identical to the 1-category polarity code of a single neuron under axonal discharge feedback, depicted in Fig. 5.1 (Baram, 2020a).

while the left neuron's self-feedback is in the "on" state. The fourth category, represented by column (d) corresponds to both neurons' self-synapses being in the "off" state.

For each of the primal-size categories of n-neuron circuits under axonal discharge self-feedback there are $C(n)$ primal-size categories under synaptic self-feedback. It follows that for circuits of 1, 2, 3 and 4 neurons there are 2, 4, 8 and 16 categories of primal sizes 2, 7, 71 and 4,111, respectively. Employing the polarity code size prime factorization numbers under axonal discharge self-feedback displayed in Table 5.1, for circuits of fiveneurons there are $32 \times 413 = 13,216$ categories of primal size 2,539, or $32 \times 17,773 = 568,736$ categories of primal size 59, or $32 \times 149,801 = 4,739,632$ categories of primal size 7. For circuits of 6 neurons there are $64 \times 4,251 = 272,064$ categories of primal size 266,239, or $64 \times 9,850,843 = 630,453,952$ categories of primal size 109, or $64 \times 29,020,051 = 1,857,283,264$ categories of primal size 7. For circuits of 7 neurons there are $128 \times 163 \times 15,451 = 322,369,664$ categories of primal size 1,746,287, or $128 \times 163 \times 1,746,287 = 36,434,531,968$ categories of primal size 15,451, or $128 \times 15,451 \times 1,746,287 = 3,453,680,695,936$ categories of primal size 163. As in the case of axonal discharge self-feedback, the three

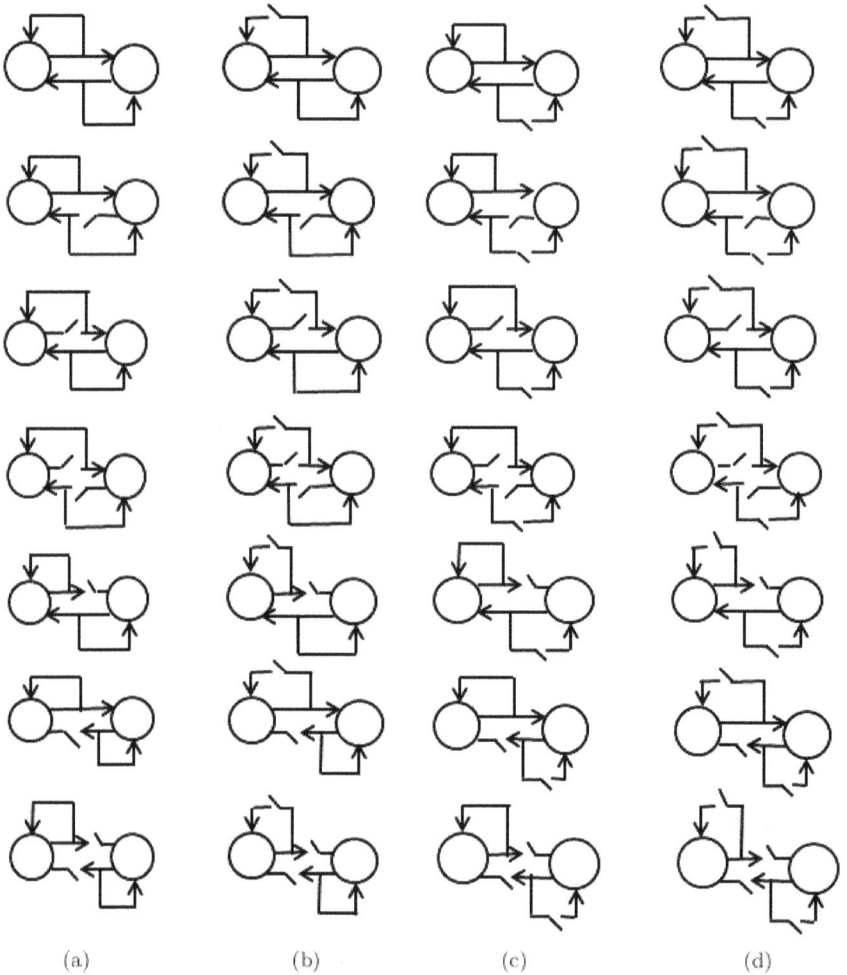

Figure 5.4. The 7-circuit categories of 2-neuron circuit polarity code represented by the four columns: (a) both self-synapses are "on", (b) left neuron's self-synapse is "off" and right neuron's self-synapse is "on", (c) left neuron's self-synapse is "on" and right neuron's self-synapse is "off" and (d) both self-synapses are "off" (Baram, 2020a).

different primal polarity category sizes obtained by factorization of the non-primal (sphenic) polarity code sizes corresponding to circuits of 5, 6 and 7 neurons may be characterized as representing three different dimensions.

The entire code of an n-neuron circuit polarity under synaptic self-feedback can be cortically represented by $Q(n)$ circuits. Consequently, the implementation of the polarity code of 1 and 2-neuron circuits would require $2 \times 2 = 4$ and $2 \times 28 = 56$ neurons, respectively. Accordingly, the implementation of the polarity codes of 3, 4 and 5-neuron circuits would require $3 \times 568 = 1,704$, $4 \times 65,776 = 263,104$ and $5 \times 39,135,393 = 167,777,120$ neurons, respectively. The implementation of the polarity code of 6-neuron circuits would require $6 \times 75,418,890,625 = 452,513,343,750$ neurons, and the implementation of the polarity code of 7-neuron circuits would require $7 \times 562,949,953,437,568 = 3,940,649,674,062,976$ neurons, far exceeding the number of neurons in the human brain(10^{12}). However, as in the case of axonal discharge self-feedback, it should be noted that any fraction of a polarity code, including such codes corresponding to circuits of n neurons where $n \geq 7$, is implementable

Table 5.2. Neural circuit polarity code and category sizes for circuits of 1–7 neurons under synaptic self-feedback. The polarity category size is primal, 2, 7, 71 and 4,111, respectively, for circuits of 1–4 neurons and sphenic for circuits of 5, 6 and 7 neurons (Baram, 2020a).

# of circuit neurons	# of self-synapse polarity permutations	Circuit polarity code size	# of primal-size categories	Primal category size
1	2	4	2	2
2	4	28	4	7
3	8	568	8	71
4	16	65,776	16	4111
5	32	33,555,424	13,216	2539
			568,736	59
			4,739,632	7
6	64	68,719,480,768	272,064	266,239
			630,453,952	109
			1,857,283,264	37
7	128	562,949,953,437,568	322,369,664	1,746,287
			36,434,531,968	15451
			3,453,680, 695,936	163

by a number of categories smaller than the one calculated for the full implementation of the corresponding polarity code.

Table 5.2 summarizes, for circuits of 1–7 neurons, the number of self-synapse polarity permutations, the circuit polarity code size, the number of primal-size polarity categories and the corresponding primal category sizes.

Finally, it might be noted that, in contrast to the polarity code addressed in Section 3.2, where connectivity is taken into account only when it is complete (without open gates), here, in order to represent complete circuit categories, incomplete connectivity (with open gates) is taken into account as well. This is evident, for instance, in the fourth row of Fig. 5.4, which, effectively, represents two isolated neurons, as does line (f) of Fig. 3.3.

5.4   Neural circuit polarity effects on firing dynamics

We now consider the effect of neural circuit polarity on its firing-rate dynamics. In particular, we examine the expression of different circuit polarity categories, implying different circuit connectivity patterns, by different firing-rate modes. We assume synaptic self-feedback, which, as explained in Section 5.3, contains axonal discharge self-feedback as a mathematical subcategory.

The neuronal firing model has evolved from the integrate-and-fire (Lapicque, 1907) and the conductance-based membrane current (Hodgkin and Huxley, 1952) paradigms, through cortical averaging (Wilson and Cowan, 1972) and neuronal decoding (Abbott, 1994) to the spiking rate model (Gerstner, 1995)

$$\tau_{m_i} \frac{d}{dt} v_i(t) = -v_i(t) + f_i \left(\boldsymbol{\omega}_i^T(t) \int_{-\infty}^{t} e^{-(t-\sigma)/\tau_i} \boldsymbol{v}_i(\sigma) d\sigma + u_i(t) \right)$$

$$(5.6)$$

where $v_i(t)$ is the firing-rate of the i'th neuron, τ_{m_i} is the corresponding membrane time constant, $\boldsymbol{v}_i(t)$ is the vector of firing-rates of the circuit's pre-neurons of the i'th neuron, ω_i is the corresponding vector of synaptic weights, τ_i is the synaptic time constant, $u_i(t)$ is the membrane activation potential, $u_i(t) = I_i(t) - r_i(t)$, with $I_i(t)$ an external input potential and $r_i(t)$ the membrane activation threshold (approximately $-60\,\text{mV}$), and where

$$f_i(x) = f(x) = \begin{cases} x & \text{if } x \geq 0 \\ 0 & \text{if } x < 0 \end{cases} \tag{5.7}$$

is the conductance-based rectification kernel (Carandini and Ferster, 2000), first observed in empirical data (Granit *et al.*, 1963; Connor and Stevens, 1971).

It has been noted that the mathematically imposing integral term in Eq. (5.6) is biologically and functionally insignificant (Wilson & Cowan, 1972), and that the essence of neural firing dynamics is captured by its instantaneous manifestation (see e.g., Dayan & Abbott, 2001; Miller and Fumarola, 2012), yielding the firing-rate model

$$\tau_{m_i} \frac{d}{dt} v_i(t) = -v_i(t) + f_i \left(\boldsymbol{\omega}_i^T(t) v_i(t) + u_i(t) \right) \tag{5.8}$$

The BCM plasticity rule (Bienenstock *et al.*, 1982), enhanced by stabilizing modifications (Intratorand Cooper, 1992; Cooper *et al.*, 2004), is a widely recognized, biologically plausible, mathematical representation of the Hebbian learning paradigm (Hebb, 1949). It suggests the computation of $\omega_i(t)$ by the model

$$\tau_{\omega_i} \frac{d}{dt} \boldsymbol{\omega}_i(t) = -\boldsymbol{\omega}_i(t) + (v_i(t) - \theta_i(t)) \, v_i^2(t) \tag{5.9}$$

where τ_{ω_i} is the learning time constant and $\theta_i(t)$ is a variable threshold satisfying (Intrator and Cooper, 1992; Cooper *et al.*, 2004)

$$\theta_i(t) = \frac{1}{\tau_{\theta_i}} \int_{-\infty}^{t} v_i^2(\tau) \exp(-(t - \tau)/\tau_{\theta_i}) d\tau \tag{5.10}$$

where τ_{θ_i} is the thresholding time constant.

The discretization of a continuous-time dynamical system (e.g. Qwakenaak and Sivan, 1972) is done under the assumption that the time steps are sufficiently small so as to allow the approximation of the input by a constant value throughout the time step. Such approximation may, in principle, introduce inaccuracies in the description of the system's behavior. Yet, discretization is often used as computational means as long as it does not introduce instability. In the present context of neural behavior, the validity of the discrete

time model can be qualitatively tested by comparison of simulations
to the ample evidence produced by neurobiological experiments. The
discrete time versions under unity time steps of Eqs. (5.8)–(5.10) are

$$v_i(k) = \alpha_i \, v_i(k-1) + \beta_i f\left(\boldsymbol{\omega}_i^T(k)v_i(k-1) + u_i\right) \qquad (5.11)$$

$$\boldsymbol{\omega}_i(k) = \varepsilon_i \boldsymbol{\omega}_i(k-1) + \gamma_i \left[v_i(k-1) - \theta_i(k-1)\right] v_i^2(k-1) \quad (5.12)$$

$$\theta_i(k) = \delta_i \sum_{i=0}^{N} \exp(-i/\tau_{\theta_i})v_i^2(k-i) \qquad (5.13)$$

where $\alpha_i = \exp(-1/\tau_{m_i})$, $\beta_i = 1 - \alpha_i$, $\varepsilon_i = \exp(-1/\tau_{\omega_i})$, $\gamma_i = 1 - \varepsilon_i$, and $\delta_i = 1/\tau_{\theta_i}$.

Since, as we have shown, circuit polarity categories grow exponentially with the number of circuit neurons, we limit our simulations
of firing-rates to circuits of two neurons.

Employing the parameter values

$$\tau_{m_1} = \tau_{m_2} = 5, \ \tau_{\omega_1} = \tau_{\omega_2} = 0.1, \ \tau_{\theta_1} = \tau_{\theta_2} = 0.1,$$

$$u_1 = u_2 = 5 \quad \text{with } v_1(0) = v_2(0) = 0 \qquad (5.14)$$

We have simulated categories (a)–(d) in Fig. 5.4. The results are
shown in Figs. 5.5–5.8, where v_1 and v_2 represent the firing-rate
sequences of the two circuit neurons in each of the cases. Each of the
pairs v_1 and v_2 should be viewed in the context of the corresponding
circuit depicted in Fig. 5.4. The details of each of the four categories
are specified in the corresponding figure captions. It can be seen
that, while the parameter values as specified by Eq. (5.14) are
identical for all neurons, the firing-rate dynamics captured by the
four categories are highly diverse (as in a natural language, the
same "words", represented here by circuit polarity, or alternatively,
by circuit firing modes, may belong to different categories).

5.5   "Working Memory" and "Magical Numbers"

As cortical information appears to be manifested by neural circuit
polarity and its consequential firing-rate, the memory capacity of
categories of small circuits, as the ones analyzed in the previous

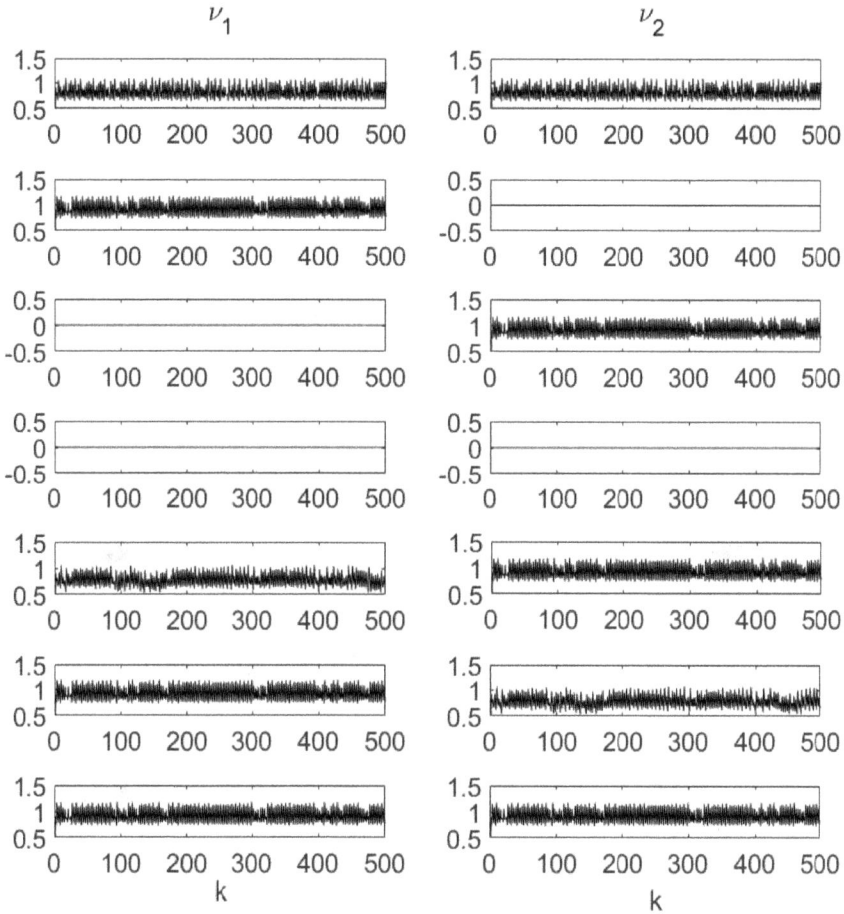

Figure 5.5. Firing-rate simulation of the 2-neuron, 7-circuit polarity category depicted in column (a) of Fig. 5.4. In the first row, the two neurons are symmetrically connected, and, consequently, their firing-rate modes are identical. In the second row, the first neuron is independently bursting, while the second is silent. In the third row, the second neuron is independently bursting, while the first is silent. In the fourth row, both neurons are silent. In the fifth and the sixth rows, while both neurons are active, their inter-neuron connections are asymmetric, and, consequently, their firing-rate modes are different. In the seventh row, the two neurons are mutually segregated by inter-neuron synapses, but, as their parameters are identical, their independent bursting firing-rate modes are identical (Baram, 2020a).

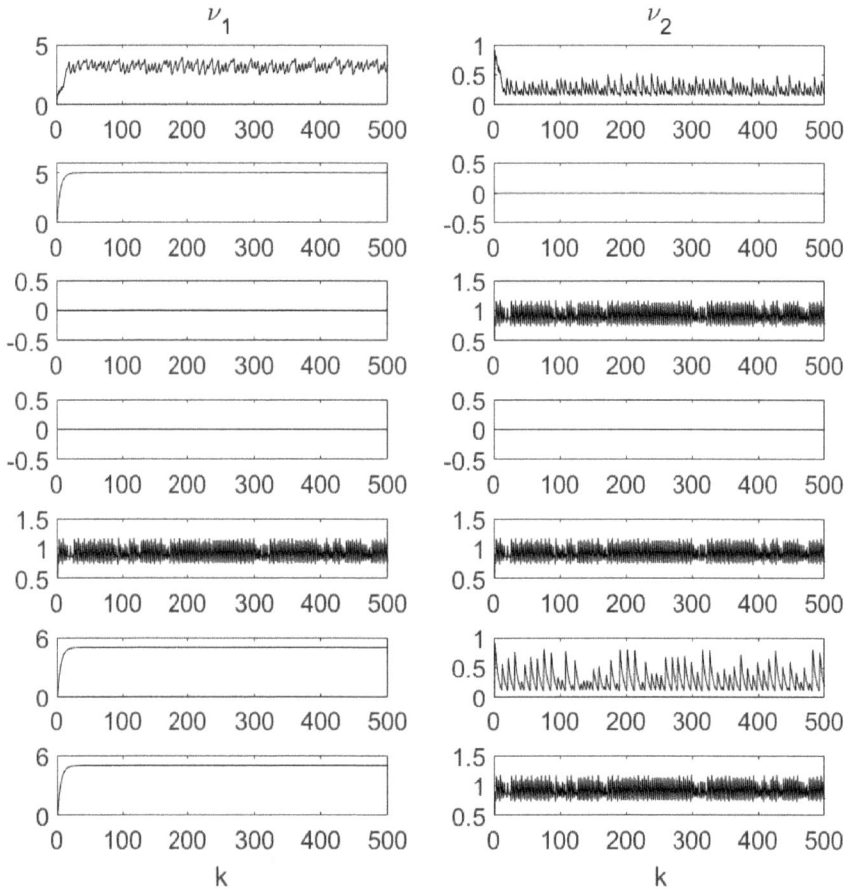

Figure 5.6. Firing-rate simulation of the 2-neuron, 7-circuit polarity category depicted in column (b) of Fig. 5.4. In the first row, as the self-synapses of the two neurons are at different polarity states, their firing-rate modes are different. In the second row, the first neuron is independently firing, while the second is silent. In the third row, the second neuron is independently firing, while the first neuron is silent. In the fourth row, both neurons have silent membranes. In the fifth and the sixth rows, differences between inter-neuron connections and self-synapses are compensated by cross-feedback and consequently, the firing-rate modes are identical. In the seventh row, the two neurons are mutually segregated, but, since their self-synapses are at different states, their firing-rate modes are different (Baram, 2020a).

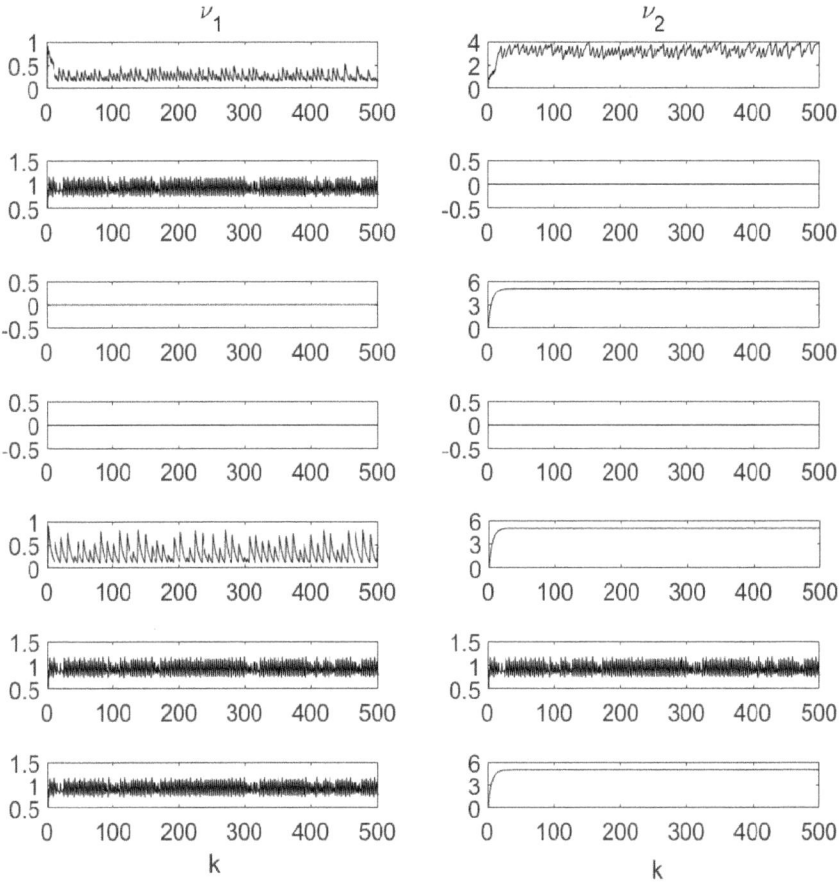

Figure 5.7. Firing-rate simulation of the 2-neuron, 7-circuit polarity category depicted in column (c) of Fig. 5.4. In the first row, the difference between the polarity states of the two self-synapses yields different firing-rate modes. In the second row, the first neuron is independently firing, while the second is silent. In the third row, the second neuron is independently firing, while the first neuron is silent. In the fourth row, both neurons have silent membranes. In the fifth and the sixth rows, differences between cross-feedback compensation of the two neurons yield different firing-rate modes. In the seventh row, the two neurons are mutually segregated, but, since their self-synapses are at different states, their firing-rate modes are different (Baram, 2020a).

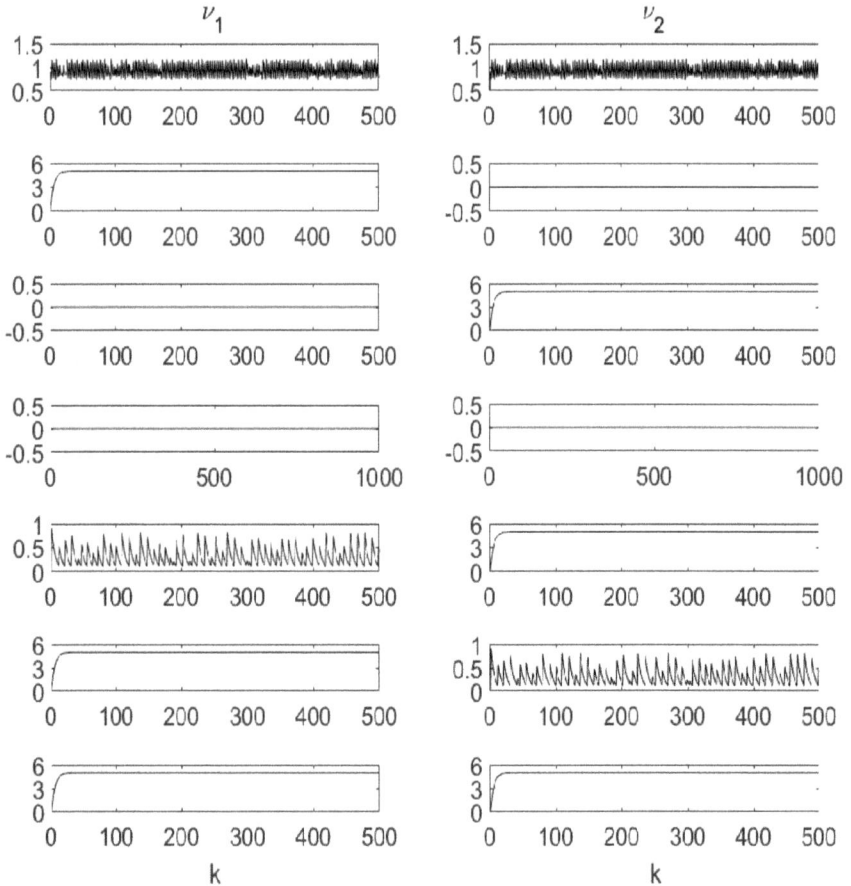

Figure 5.8. Firing-rate simulation of the 2-neuron, 7-circuit polarity category depicted in column (d) of Fig. 5.4. In the first row, the identical polarity states of the two self-synapses, and the identical polarity states of the two cross-neuron synapses yield identical firing-rate modes for the two neurons. In the second row, the first neuron is independently firing, while the second has a silent membrane. In the third row, the second neuron is independently firing, while the first neuron has a silent membrane. In the fourth row, both neurons have silent membranes. In the fifth and the sixth rows, differences between cross-feedback compensation of the two neurons yield different firing-rate modes. In the seventh row, the two neurons are mutually segregated by silent inter-neuron synapses, and, having both their self-synapses silent, fire in identical firing-rate modes (Baram, 2020a).

sections, is of interest. Employing experimental behavior data, it has been suggested that human capacity for information processing is limited by "the magical number seven, plus or minus two" (Miller, 1956). Pertaining to "working memory" (Pribram *et al.*, 1960) or metaphorically, "blackboard memory" (Colman, 2008) and later put on molecular basis (Mongillo *et al.*, 2008) and supported by a winnerless competition dynamical model (Bick and Rabinovich, 2009), this limit has been associated with "one dimensional" objects (such as musical tunes, electrical skin stimulation or visual dot location) while noting that "higher dimensional" objects (such as human faces) would require higher information processing capacities (Miller, 1956). As we have noted in the present work, dimensionality of information items may be associated with prime factorization of composite polarity code-sizes. Specifically, in contrast to the single primal polarity code sizes corresponding to circuits of 1–4 neurons, which may be characterized as having a single dimension, the three different primal polarity category sizes obtained by factorization of *sphenic* (Lehmer, 1936) polarity code sizes corresponding to circuits of 5 and 6 neurons may be characterized as representing three different dimensions. Another proposition has suggested "the magical number 4" (or, more specifically, "between 3 and 5", Cowan, 2001) instead of "the magical number 7" (Miller, 1956). We note that, as seen in Table 5.2, the number 7 appears as the polarity category size for a circuit of two neurons, while the number 4 appears as the polarity code size for a single neuron. This may suggest that the two numbers, 7 and 4, have been experimentally measured in two different contexts, one of a single category of items, represented by 2-neuron circuits with axonal discharge self-feedback, and the other of two different categories of items, represented by 1-neuron circuits with synaptic self-feedback. A division of memory into separate domains of small and large numbers of information items, as suggested by the large spread of polarity category and code sizes represented by Table 5.2, has also been suggested on experimental grounds (Cowan, 1995). Indeed, outstanding processing abilities of many information items have been experimentally noted (Chase and Ericsson, 1982; Yaro and Ward, 2007; Neumann, 2010).

A question of interest is whether there is a mathematical basis to "magical numbers" as memory capacities. A widely recognized concept of memory in the context of binary neuronal states is the Hebbian learning paradigm (Hebb, 1949), which has been formalized as follows (McCulloch and Pitts, 1943; Hopfield, 1982). Consider a neural circuit of n neurons, whose membrane and synapse polarity states, $w_j(i) \in (0, 1)$, $j = 1, \ldots, n$ at discrete times, $i = 1, \ldots, M$, take the values 0 or 1 with equal probabilities. Storage of polarity state values is performed according to the rule

$$\boldsymbol{\omega}_{k,j} = \sum_{i}^{M} \omega_k^{(i)} \omega_j^{(i)} \tag{5.15}$$

Retrieval is performed by the rule

$$x_k = \begin{cases} 0, & \text{if } u_k \leq t \\ 1, & \text{if } u_k > t \end{cases} \tag{5.16}$$

where t is a threshold and

$$u_k = \sum_{j=1}^{n} \boldsymbol{\omega}_{k,j} z_j \tag{5.17}$$

is the input to the k'th neuron, where z_j, $j = 1, \ldots, n$ are the components of a probe.

The memory capacity of binary $(0,1)$ vectors of size n by networks of the type Eqs. (5.15)–(5.17) has been shown to be about $0.14n$ (Amit *et al.*, 1987). We note that, from a mathematical viewpoint, n may represent the number of circuit neurons, in one context, and the number of circuit polarity gates, in another. For polarity vectors of size 7, associated, as shown in the present work, with 2-neuron circuits, this result would imply a Hebbian memory capacity, by Eqs. (5.15)–(5.17), of approximately one polarity vector per circuit $(0.14 \times 7 = 0.98)$. A "working memory" consisting of 2-neuron circuits would allow for a maximum of 7 circuits having different polarity states, in each of the categories depicted in Fig. 5.4. With each of these circuits storing one polarity state vector, this would establish a neural embodiment of "the magical number 7" as the category memory capacity.

5.6   Discussion

A neural circuit polarity code is defined by the complete set of circuits generated by permutations of polarity states of the circuit neurons' membranes and synapses. For a given number of circuit neurons, a primal circuit polarity code size implies that the code is indivisible, constituting a circuit polarity category. The present chapter has shown that when the circuit polarity code size is not primal, it can be factorized into primal-size subcodes, representing different categories, which, in turn, produce different firing-rate modes. Under axonal self-feedback, fully connected circuits of 1, 2, 3 and 4 neurons produce circuit polarity codes of primal sizes 2, 7, 71 and 4,111, respectively. On the other hand, circuits of 5, 6 or 7 neurons produce circuit polarity codes of composite sizes. Such codes can be computationally transformed into primal-size polarity categories by prime factorization of their polarity code sizes. The high dominance of the first term on the right-hand-side of Eq. (5.3) ($2^{n(n-1)}$) implies that such prime factorization is practically impossible by standard techniques for circuits of more than 7 neurons. It is conceivable that nature would have its way of producing primal-size polarity categories for circuits of more than 7 neurons. Yet, even without such capabilities, relatively large cortical areas may be aroused by smaller circuits interacting in a feed-forward fashion.

The effect of neuronal self-feedback on primal polarity code size is particularly interesting in the context of a long-standing debate concerning the nature of such feedback. While, employing neurophysiological and molecular considerations, the synaptically mediated self-feedback paradigm (Groves *et al.*, 1975) has been presented as an alternative to the axonal discharge self-feedback paradigm (Carlsson and Lindquist, 1963; Smith and Jahr, 2002), we have shown in this work that the first paradigm contains the second. Specifically, while the axonal discharge self-feedback paradigm defines a single, primal-size, circuit polarity category for certain fully connected circuit sizes

(1–4), the synaptic self-feedback paradigm defines the entire range of such categories produced by permutation of self-synaptic polarity states. Consequently, the number of categories produced by the synaptically mediated self-feedback paradigm grows exponentially in the number of circuit neurons with respect to that produced by the axonal discharge self-feedback paradigm.

"Magical numbers", supported by psychophysical tests of "absolute judgement" (Miller, 1956), have been suggested as limit capacities for "working memory" (Pribram *et al.*, 1960). While such numbers were not directly related to neurophysiological experiments, we have suggested that their small values (3–9 according to various publications) are resembled by the small circuit sizes suggested in the present study as the "alphabets", "words" or "categories", of cortical information. We have also suggested that the level of prime factorization of circuit sizes constitutes a certain notion of informational dimension. The possible neurophysiological basis for such correspondence is proposed for future research.

Finally, we note that the practical advantage of primality may be explained by the ability of different prime numbers to generate large numbers of different combinations without repetition. This property has been characterized in purely mathematical contexts (Furstenberg, 1955) and further in the geometric contexts of polygon construction and partition (Křížek *et al.*, 2001). Its applicability in digital computation has been noted in the contexts of hash table generation (Cormen *et al.*, 2001), error detection (Kirtland, 2001) and pseudo-random number generation (Matsumoto and Nishimura, 1998). Prime numbers have been shown to play fundamental roles in quantum physics (Peterson, 1999) and in quantum computation (Bengtsson and Życzkowski, 2017) as well as the natural survival game of predator and prey (Williams and Simon, 1995), literature (Ribenboim, 2017) and music (du Sautoy, 2003). Yet, it appears that the relevance of prime numbers in cortical contexts has not been noted before.

Chapter 6

Primal-size Neural Circuits in Trees of Meta-periodic Interaction

6.1 Introduction

Early studies of information representation in neural circuits have addressed binary $(0, 1)$ neuronal states of silence and activity (McCulloch and Pitts, 1943; Amari, 1972; Hopfield, 1982). Studies of information storage and retrieval capacities have generally suggested, explicitly or implicitly, a direct linkage between high capacity and large circuit size (McEliece *et al.*, 1987; Amit *et al.*, 1987; Baram and Sal'ee, 1992). However, the focus on a large number of neurons has overlooked the segregated nature of cortical information and functionality. As, in addition, the binary setting did not appear to have been directly motivated by neurophysiological consideration at the time, it has largely remained in the realms of information theory, mathematics and physics, essentially disregarded in biological circles. Later experimental neurophysiological findings, concerning somatic (Melnick, 1994) and synaptic (Atwood and Wojtowicz, 1999) silencing and reactivation, have been termed "neural circuit polarity" (Baram, 2018). Cortical activity segregation and integration have been argued on grounds of thalamocortical simulations (Stratton and Wiles, 2015). A realization of the significant effects of neuronal circuit polarity and polarity-induced segregation on cortical connectivity, firing dynamics and memory in recent studies (Baram, 2018) has motivated the present examination of random circuit segregation by combining two mathematical contexts. The first is random graph theory, which reveals rather unexpected, yet highly powerful results relevant to circuit connectivity. The second is the theory of prime numbers, which, intriguing mathematicians for several centuries, is increasingly finding its way into the natural sciences. While prime numbers have been recently shown to be central to neural circuit polarity categorization (Baram, 2020a), their key role in cortical meta-period maximization appears to have been noted only recently (Baram, 2021). While individual cyclic circuits, controlled by inhibitory inter-neuron potentiation (Markram *et al.*, 2004), appear to analytically ratify experimental findings on relatively low working memory capacities of the human brain (Miller, 1956; Pribram *et al.*, 1960), the concept of meta-periodicity, facilitated by excitatory inter-circuit synapses (Battaglia *et al.*, 2012), is shown here to extend to trees of primal size neural circuits, producing large non-repetitious cortical memory capacities.

6.2   Primal-size neural circuits in meta-periodic interaction

A correspondence between neural circuits and random graphs becomes self-evident, when nodes are represented by neurons, and edges are represented by biologically supported positive membrane (Melnick, 1994) and synapse (Atwood and Wojtowicz, 1999). Viewing *connected neural circuits* as *connected components* in the sense defined in previous chapters, a question of interest is whether results from the theory of random graphs offer potential benefits in the understanding of cortical properties. As we show next, the answer becomes markedly positive when an additional mathematical discipline, namely, number theory, is brought into play. More specifically, prime numbers, having attracted the interest of mathematicians for several centuries, and revealed in a variety of theoretical and experimental scientific disciplines, is found to play a major role in maximizing the information capacities of cortical structures.

According to Eqs. (2.3) and (2.4), large circuits tend to be connected with high probability. A connected neural circuit represents a single cortical word. For a large neural circuit, complete connectivity would mean that the entire circuit represents a single word.

Large neural circuit segregation into smaller subcircuits becomes an obvious necessity for reasonably short words in cortical language. An average of 7×10^3 synapses per neuron (Drachman, 2005) in a human brain of 10^{11} neurons (von Bartheld *et al.*, 2016) would yield an average edge probability of $p = 7000/10^{22} < 10^{-18} < 10^{-11} = 1/n$, which satisfies the subcriticality condition for the satisfaction of Eq. (2.5). As will be demonstrated in the sequel, segregated circuits of primal sizes produce the longest sequential meta-periods without repetitions. Yet, they maintain the flexibility of connecting in different arrangements, producing different meta-words.

The capacity of neural circuit segregation is essentially defined by the graph theoretic results noted in Section 2.2. While the largest connected circuit for $n = 10^{11}$ neurons and for $p < 1/n$ (sub-criticality) has $[\log(10^{11})] = 25$ neurons, the largest such circuit whose size is a prime number has 23 neurons (neural assemblies of size $n = 10^{12}$ or smaller do not produce primal values greater than 23 for log(n). The next prime, 29, arises for $n > 10^{12}$, as $\log(10^{12}) < 29 < \log(10^{13})$). Table 6.1 displays all the prime numbers S between 2 and 23, inclusive. Each of these prime numbers is coupled with a number $[n(S)]$ which represents the nearest decimal power greater than $n(S) = \exp(S)$.

It might be noted that the requirement $n \to \infty$ associated with Eq. (2.5) suggests that only high values of n may be considered in the present context. Yet, the question remains, how high is "high". There does not appear to be a good answer to this question. The requirement $n \to \infty$, driven by mathematical rationale in the underlying graph theory, does not appear to invalidate the small-circuit results. As demonstrated by Table 6.1, the small-circuit results are in close agreement with the large-circuit ones, regarding order of magnitude.

Figure 6.1(a) shows two primal-size circuits in sequence, one of 2 neurons, represented by circles and denoted 1 and 2, and the other of 3 neurons, denoted a, b, and c. Inter-neurons in the central nervous system are primarily inhibitory (Markram *et al.*, 2004), which is also supported by the large majority of inhibitory cortical connectivity (Mongillo *et al.*, 2018).

Table 6.1. Maximal primal sizes S of connected circuits for different numbers of neuron assemblies n in the relevant cortical range $n = 10^1$ to $n = 10^{10}$. $[n(S)]$ represents the nearest decimal power greater than $n(S) = \exp(S)$ (yielding $S = \log(n(S))$), in the sense of Eq. (2.5) (Baram, 2021).

S	2	3	5	7	11	13	17	19	23
$[n(S)]$	10^1	10^2	10^3	10^4	10^5	10^6	10^7	10^8	10^{10}

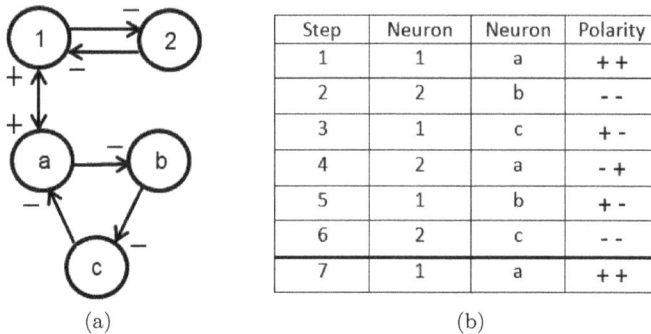

Step	Neuron	Neuron	Polarity
1	1	a	+ +
2	2	b	- -
3	1	c	+ -
4	2	a	- +
5	1	b	+ -
6	2	c	- -
7	1	a	+ +

(a)　　　　　　　　　　　　　　　　(b)

Figure 6.1. (a) Segregated primal size neural circuits of 2 and 3 neurons, having (b) A meta-period of 6 steps (on the 7'th step the meta-period starts all over). Internal circuit polarities are negative (inhibitory) while inter-circuit polarities are positive (excitatory). A meta-period is initiated when both interacting neurons, 1 and a, are active (Baram, 2021).

The cyclic connectivity implies that a neuron will only fire for a certain period once the neuron preceding it in the cycle has stopped firing (called Post-inhibitory Facilitation (Dodla *et al.*, 2006), or Post-inhibitory Rebound (Jones and Thompson, 2001)). Thus inhibition, or negative polarity, is followed by positive polarity, which implies a brief firing period. On the other hand, the interaction between neurons of different circuits and different cortical areas is primarily excitatory and simultaneous (Battaglia *et al.*, 2012).

Neuron 1 and neuron a of the corresponding circuits in Fig. 6.1(a) are connected bi-directionally (which also represents simultaneous mutual activation of the two neurons), while the other neurons of each of the two circuits are connected directionally in a cycle.

The inhibitory (negative polarity) nature of the inter-neurons in each of the two circuits is represented by minus signs, while the excitatory (positive polarity) nature of the inter-circuit neurons 1 and a is represented by plus signs.

Figure 6.1(b) presents the 6-step meta-periodic cycle of the two circuits. It can be seen that the 2-neuron circuit alternates between neurons 1 and 2, in the order $1 \rightarrow 2 \rightarrow 1 \rightarrow 2 \rightarrow 1 \rightarrow 2$ while the 3-neuron circuit undergoes two cycles of 3-steps each $a \rightarrow b \rightarrow c \rightarrow a \rightarrow b \rightarrow c$, so that neuron 1 and neuron a fire together again at step 7, so as to start a new 6-step meta-period. The polarities of the corresponding neurons in each of the steps, specified in the last column, indicate the present state in the meta-period, which starts and ends when both neurons 1 and a are in simultaneous positive polarity states $(++)$.

Figure 6.2(a) shows two circuits, one, as in Fig. 6.1, of 2 neurons denoted 1 and 2, and the other of 4 neurons, denoted a, b, c and d. The main difference between this case and that depicted in Fig. 6.1(a) is that, in contrast to the primal circuit size of 3 shown in Fig. 6.1, the circuit size of 4 depicted in Fig. 6.2(a) is not primal. As in Fig. 6.1(a), it can be seen that neuron 1 and neuron a of the corresponding circuits are connected bi-directionally (which represents simultaneous mutual activation by the two neurons), while the other

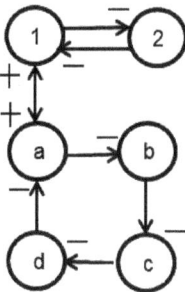

Step	Neuron	Neuron	Polarity
1	1	a	+ +
2	2	b	- -
3	1	c	+ -
4	2	d	- -
5	1	a	+ +

(a)   (b)

Figure 6.2.   (a) Segregated neural circuits of 2 and 4 neurons, having (b) A meta-period of 4 steps (on the 5'th step the meta-period starts all over). Internal circuit polarities are negative (inhibitory) while inter-circuit polarities are positive (excitatory, Baram, 2021).

neurons of each of the two circuits are connected directionally in a cycle. As in the previous case, the inhibitory (negative polarity) nature of the inter-neurons in each of the two circuits is represented by minus signs, while the simultaneous excitatory (positive polarity) states of the inter-circuit neurons 1 and a is represented by plus signs. Figure 6.2(b) presents the 4-step meta-periodic cycle of the two circuits. It can be seen that the 2-neuron circuit alternate between neurons 1 and 2, in the order $1 \to 2 \to 1 \to 2$ while the 4-neuron circuit undergoes one cycle of 4-steps $a \to b \to c \to d$ so that the two circuits meet again at step 5, so as to start a new 4-step meta-period. The polarities of the corresponding neurons in each of the steps, specified in the last column, indicate the present state of the meta-period, which starts and ends when both neurons 1 and a are in simultaneous positive polarity states $(++)$.

It follows that while the 2-circuit arrangement in Fig. 6.2(a) involves more neurons (6) than the one in Fig. 6.1(a) (5), the information capacity of the latter, which produces a longer (6 steps) non-repetitive meta-period, is greater than that of the former (4 steps). This property of prime numbers, generating longer indivisible meta-periods than cycles involving larger neighboring non-primal numbers, extends to all primes.

6.3 Primal-size circuit interaction in firing-rate meta-periodicity

Neural circuit connectivity is only one part of the cortical language. The other, closely related, part is the neural firing-rate dynamics. The instantaneous discrete-time firing-rate model for the individual neuron is (e.g., Baram, 2018)

$$v_i(k) = \alpha_i \, v_i(k-1) + \beta_i f_i(\boldsymbol{\omega}_i^T(k)\boldsymbol{v}_i(k-1) + u_i) \qquad (6.1)$$

where $\boldsymbol{v}_i(k)$ is the vector of the i'th neuron's pre-neurons' firing-rates (including self-feedback), v_j, $j = 1, 2, \ldots, n$, $\alpha_i = \exp(-1/\tau_{v_i})$ and $\beta_i = 1 - \alpha_i$, with τ_{v_i} the membrane time constant of the i'th neuron, $\boldsymbol{\omega}_i(k)$ is the vector of synaptic weights corresponding to the i'th neuron's pre-neurons (including self-feedback), $u_i = I_i - r_i$, with I_i the circuit-external activation input and r_i the membrane resting potential of the i'th neuron, and f_i is the conductance-based rectification kernel defined as (Carandini and Ferster, 2000)

$$f_i(x) = \begin{cases} x & \text{if } x \geq 0 \\ 0 & \text{if } x < 0 \end{cases} \tag{6.2}$$

The discrete-time Bienenstock-Cooper-Munro plasticity rule (Bienenstock *et al.*, 1982) takes the multi-neuron form

$$\boldsymbol{\omega}_i(k) = \varepsilon_i \boldsymbol{\omega}_i(k-1) + \gamma_i [v_i(k-1) - \theta_i(k-1)] \boldsymbol{v}_i^2(k-1) \tag{6.3}$$

where $\boldsymbol{v}_i^2$ is the vector whose components are the squares of the components of $\boldsymbol{v}_i$, and

$$\theta_i(k) = \delta_i \sum_{\ell=0}^{N} \exp(-\ell/\tau_{\theta_i}) v_i^2(k-\ell) \tag{6.4}$$

with $\varepsilon_i = \exp(-1/\tau_{\omega_i})$, $\gamma_i = 1 - \varepsilon_i$ and $\delta_i = 1/\tau_{\theta_i}$, where τ_{ω_i} and τ_{θ_i} are the corresponding time constants.

Next, we analyze and demonstrate the effects of primal-size circuit interaction on the joint firing-rate dynamics. The essence of these findings will be demonstrated for small primal-size circuits of 2 and 3 neurons, as depicted in Fig. 6.1(a), since the simulation of larger primal size circuit firing would be highly elaborate and tedious.

Let the neurons in the 2-neuron circuit of Fig. 6.1(a) have the parameters:

$$\tau_{v_1} = 2,\ \tau_{\omega_1} = 300,\ \tau_{\theta_1} = 0.1,\ u_1 = 5,\ \theta_1(0) = \omega_1(0) = 0,\ v_1(0) = 1$$

$$\tau_{v_2} = 1,\ \tau_{\omega_2} = 1,\ \tau_{\theta_2} = 0.1,\ u_2 = 5,\ \theta_2(0) = \omega_2(0) = 0,\ v_2(0) = 1$$

and let the neurons in the 3-neuron circuit in Fig. 6.1(a) have the parameters

$$\tau_{v_a} = 1,\ \tau_{\omega_a} = 0.7,\ \tau_{\theta_a} = 0.5,\ u_a = 3,\ \theta_a(0) = \omega_a(0) = 0,\ v_a(0) = 1$$

$$\tau_{v_b} = 2,\ \tau_{\omega_b} = 300,\ \tau_{\theta_b} = 0.1,\ u_b = 3,\ \theta_b(0) = \omega_b(0) = 0,\ v_b(0) = 1$$

$$\tau_{v_c} = 2,\ \tau_{\omega_c} = 1,\ \tau_{\theta_c} = 0.1,\ u_c = 3,\ \theta_c(0) = \omega_c(0) = 0,\ v_c(0) = 1$$

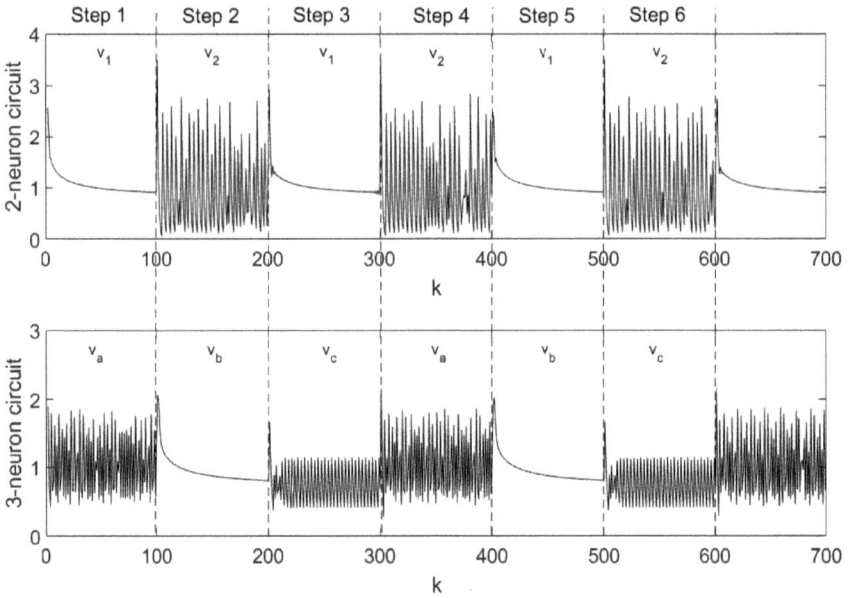

Figure 6.3. 6-step meta-period firing-rate dynamics of the combined circuit comprising the primal-size circuits of 2 and 3 neurons depicted in Fig. 6.1(a) (Baram, 2021).

The two primal-size circuits have a joint period of 6 steps, as specified in Fig. 6.1(b). The corresponding neuronal firing-rate dynamics are depicted in Fig. 6.3. It can be seen that, while the 2-neuron circuit maintains its 2-step alternating convergent-tonic (v_1) and chaotic (v_2) firing-rate modes in the two neurons, the 3-neuron circuit keeps 3-step dynamic modes: chaotic (v_a), convergent-tonic (v_b) and oscillatory (v_c).

6.4   Cortical information capacities under primal-size neural circuits

Given an assembly of n neurons with connectivity probability $p = c/n$ with $c < 1$ (the subcriticality condition), Eq. (2.5) implies a minimal number of

$$C(n) = \frac{n}{\log(n)} \tag{6.5}$$

connected neural circuits. Associating such circuits with code words of cortical information, Eq. (6.5) would constitute, then, a lower bound on the information capacity (in number of words) of n such sparsely connected neurons. n may take different values, corresponding to different cortical structures. Specifically, we have shown that even for the large number of neurons in the human brain, random graph theory, combined with mathematical number theory, puts a relatively low upper limit (23) on the primal neural circuit size, while facilitating, as we show next, a high information capacity. In this context, capacity would correspond to the number of neural circuits which are internally connected in the graph-theoretic sense. The total number of such circuits may change according to circuit polarity changes. The assignment of neural circuits to different cortical structures would determine circuit sizes according to their location and functionality.

A particularly unique aspect of primal-size neural circuits presented in this work in the context of cortical information is the concept of meta-periodicity, which, as we have shown in the previous section, involve both time and firing-rate dynamic modes as expressions of cortical information. Memory is not an instantaneous event. Stored in the form of circuit polarities, memory is retrieved as a meta-period of firing-rate modes, generated by circuit interaction. Memory capacity can therefore be defined on two additional separate, yet related levels: the maximal number and the maximal length of the meta-periods that can be cortically stored and retrieved.

Table 6.2 specifies the lengths of meta-periods corresponding to interactions between circuits of sequential primal sizes in the relevant range 1–23. While mathematical number theory excludes 1 as a prime number, graph-theory counts single isolated nodes as "components". Similarly, single isolated neurons, having characteristic firing-rate modes, are justifiably counted as isolated neural circuit. Yet, it can be seen that such inclusion does not change the results of the table.

As shown in Table 6.2, the meta-periodic length in this range grows from 1 to 223,092,870. Products of any subgroup of the primes involved represent valid meta-periods as well. As each meta-period

Table 6.2. Meta-period lengths of sequentially interacting primal-size circuits in the range 1–23 (Baram, 2021).

1
2
1x2x3=6
1x2x3x5=30
1x2x3x5x7=210
1x2x3x5x7x11=2,310
1x2x3x5x7x11x13=30,030
1x2x3x5x7x11x13x17=510,510
1x2x3x5x7x11x13x17x19=9,699,690
1x2x3x5x7x11x13x17x19x23=223,092,870

represents an information item, the maximal length the meta-periods that can be generated represents, in a sense, a cortical information capacity. As was illustrated by Figs. 6.1 and 6.2, and applies, by virtue of primality, to the general case, replacing any primal number by a neighboring greater non-primal number will reduce the lengths of indivisible meta-periods.

More generally, the number of indivisible meta-periods that can be composed and stored (by circuit polarities) and retrieved (by meta-periodic firing-rates) can be derived by rather simple calculations. First we note that the maximal number of selections of different primal circuit sizes in the range 1–23 is

$$\sum_{k=1}^{9} \binom{9}{k} = 2^9 - 1 = 511 \qquad (6.6)$$

(as explained above, circuits of size 1 are also included for the present purposes). 10 such selections are shown in the 10 rows of Table 6.2. Yet, in the context of cyclic circuit activity, given a k-neuron circuit there are $k!$ (k-factorial) ways to select the cyclic order of firing in the circuit neurons and its manifestation would require $k! \times k$ neurons.

It follows that there are $23! = 2.5852 \times 10^{22}$ possible cyclic arrangements of a connected circuit of 23 neurons, with each such circuit representing a word in cortical language. The neural circuit representation of all of these words would require $23 \times 23! = 5.9460 \times 10^{23}$, far exceeding the number of neurons in the human brain. The last row of Table 6.2 displays a single meta-period, composed of circuits of primal sizes 1–23 (the formally non-primal 1 included, as explained above). The cortical manifestation of just a single such meta-period would require, as specified in the table, 223,092,870 neurons. The cortical manifestation of the entire variety of such meta-periods, generated by order permutations in the different circuits involved is impossible, as the entire size of the human brain would be exhausted by just a few order permutations.

In order to proceed with the search for cortical meta-period capacity, let us consider connected neural circuits of primal size 11 or less. The number of meta-periods that can be numerically generated in this case is

$$M = 2! \times 3! \times 5! \times 7! \times 11! = 2.897 \times 10^{14} \tag{6.7}$$

requiring

$$N = (2 \times 2!) \times (3 \times 3!) \times (5 \times 5!) \times (7 \times 7!) \times (11 \times 11!)$$
$$= 6.6921 \times 10^{17} \tag{6.8}$$

neurons, yet, far exceeding the total estimated number of neurons in the human brain.

Further reducing the maximal primal neural circuit size to 7 neurons (which would imply circuit sizes 1, 2, 3, 5 and 7) produces a variety of

$$M = 2! \times 3! \times 5! \times 7! = 7,257,600 \tag{6.9}$$

meta-periods, requiring

$$N = (2 \times 2!) \times (3 \times 3!) \times (5 \times 5!) \times (7 \times 7!) = 1.5241 \times 10^9 \tag{6.10}$$

neurons for implementation. Nearing the total estimated number of neurons in the human brain (8.6×10^{10}, von Bartheld *et al.*, 2016), the cortical implementation of the entire variety of meta-periods composed of interacting circuits of primal size 7 or less

consumes most of the capacity of the human brain, with some room
to spare. The remaining neurons could accommodate longer meta-
period, comprising circuits of larger primal sizes (limited by 23).

Finally, it might be noted that working memory capacity greater
than "seven, plus or minus two" has been ruled out on experimental
grounds (Miller, 1956; Pribram *et al.*, 1960). Exceptional brains
of higher capacities have been experimentally noted (Chase and
Ericsson, 1982; Yaro and Ward, 2007; Neumann, 2010). In the present
context, memory is manifested by meta-periods which may associate
each memory item with a firing-rate mode. A cyclic circuit of seven
different neurons would sequentially produce seven different firing-
rate modes which may represent seven memory items. As shown
in this work, such cycles may be combined with shorter primal-
length cycles generated by smaller primal-size circuits, producing
more composite memory items.

6.5   Primal-size neural circuits in meta-periodic trees

Small primal-size circuits may combine so as to construct large
information structures in a tree-like fashion. As musical notes
corresponding to auditory frequencies constitute "one-dimensional"
information items, we employ a musical example as means for
conveying the underlying concept of primal-size neural circuit trees.
This will extend the concept of meta-periodic sequence construction
by two interacting circuits addressed in previous sections to neural
circuit trees of any size.

Consider a musical piece played jointly by three musicians, one
on a cello, one on a viola and one on a violin. Each of the instruments
produces a different range of auditory frequencies, which are cortically picked up by different types of neurons, arranged in separate
circuits. As before, while the circuits are essentially segregated, there
is a single inter-circuit excitatory connection between consecutive
layers of the tree.

Suppose that the cello plays a repeated sequence of two tunes,
the viola plays a repeated sequence of three tunes and the violin plays

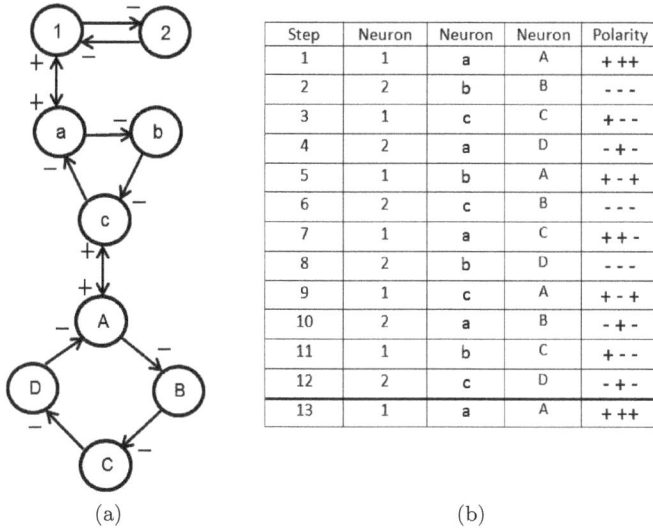

Step	Neuron	Neuron	Neuron	Polarity
1	1	a	A	+ ++
2	2	b	B	- - -
3	1	c	C	+ - -
4	2	a	D	- + -
5	1	b	A	+ - +
6	2	c	B	- - -
7	1	a	C	+ + -
8	2	b	D	- - -
9	1	c	A	+ - +
10	2	a	B	- + -
11	1	b	C	+ - -
12	2	c	D	- + -
13	1	a	A	+ ++

(a)   (b)

Figure 6.4. (a) A 3-layer tree structure of 2, 3 and 4-neuron circuits having (b) A meta-period of 12 steps (on the 13'th step the meta-period starts all over). Internal circuit polarities are negative (inhibitory) while inter-circuit polarities are positive (excitatory, Baram, 2021).

a repeated sequence of four tunes. The tunes played by each of the instruments are cortically picked up by neural circuits consisting of 2, 3 and 4 neurons of the corresponding types respectively. The three circuits are first arranged in the tree structure depicted by Fig. 6.4. It can be seen in Fig. 6.4(a) that, neuron 1 and neuron a of the first and the second circuits, respectively, and neuron c and neuron A of the second and third circuit, respectively, are connected bi-directionally (which also represents simultaneous mutual activation of each of the two pairs of neurons), while the other neurons of each of the three circuits are connected directionally in a cycle. The inhibitory (negative polarity) nature of the interneurons in each of the three circuits is represented by minus signs, while the excitatory (positive polarity) nature of the inter-circuit neurons 1, a and A is represented by plus signs.

Figure 6.4(b) presents the 12-step meta-periodic cycle of the three circuits. It can be seen that the 2-neuron circuit alternates between neurons 1 and 2, in the order $1 \to 2 \to 1 \to 2 \to 1 \to 2$ while the 3-neuron circuit undergoes two cycles of 3-steps each, $a \to b \to c \to a \to b \to c$, and the 4-neuron circuit undergoes three 4-step cycles, $A \to B \to C \to D \to A \to B \to C \to D \to A \to B \to C \to D$, so that neurons 1, a and A fire together again at step 13, so as to start a new 12-step meta-period. The polarities of the corresponding neurons in each of the steps, specified in the last column, indicate the present state in the meta-period, which starts and ends when neurons 1, a and A are in simultaneous positive polarity states $(+ + +)$.

Next suppose that, while the cello and the viola continue to play the 2-tune and the 3-tune sequence as before, the violin plays a 5-tune sequence, instead of the 4-tune sequence it played before. The three corresponding neural circuits are depicted in Fig. 6.5(a). Figure 6.5(b) represents the 30-step meta-periodic cycle of the three

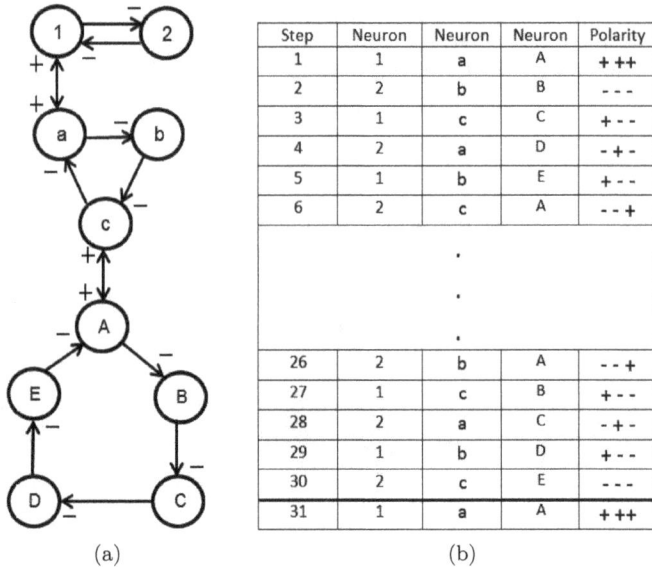

Step	Neuron	Neuron	Neuron	Polarity
1	1	a	A	+ ++
2	2	b	B	- - -
3	1	c	C	+ - -
4	2	a	D	- + -
5	1	b	E	+ - -
6	2	c	A	- - +
		.		
		.		
		.		
26	2	b	A	- - +
27	1	c	B	+ - -
28	2	a	C	- + -
29	1	b	D	+ - -
30	2	c	E	- - -
31	1	a	A	+ ++

(a)   (b)

Figure 6.5. (a) A 3-layer tree structure of 2, 3 and 5-neuron circuits having (b) A meta-period of 30 steps (on the 31'st step the meta-period starts all over). Internal circuit polarities are negative (inhibitory) while inter-circuit polarities are positive (excitatory, Baram, 2021).

circuits. The change from a meta-periodic sequence of 12 steps (Fig. 6.4(b)) to a meta-periodic sequence of 30 steps (Fig. 6.5(b)) is not due to the fact that 5 is greater than 4. To see this, suppose that the violin's repeated sequence is increased in length from 5 to 6 tunes. The length of the joint meta-periodic sequence will drop back from 30 to 12 steps. On the other hand, an increase in the length of the violin's repeated sequence from 6 to 7 tunes will increase the joint meta-periodic sequence to 42 steps. The "secret" is obviously in the difference between the primal nature of the numbers 5 and 7, on the one hand, and the non-primal nature of the numbers 4 and 6, on the other.

As words are of higher dimension than letters and musical notes (which can each be represented by neurons), the cortical representation of words of natural language would seem to require more elaborate "hardware". Specifically, each word would require a neural circuit of appropriate size. The tree structure of a sentence would be conceptually similar to the structures depicted in Figs. 6.4 and 6.5, with essentially two differences:

1) Neurons would be replaced by whole circuits, each representing a word.
2) The readout of each circuit would consist of a single cycle which would reveal the word it represents.

6.6 Discussion

Employing fundamental results from random graph theory and the meta-periodicity of interacting primal-size neural circuits, we have shown that small neural circuits offer high capacities for information representation, manipulation and memory. Specifically, while large assemblies of neurons, forming random graphs, tend to connect into large circuits, representing long words, circuits of small primal size, representing short words, may still produce very long meta-periodic structures of information, which may span different cortical areas of different circuit types and functionality. Similar to natural and artificial languages, such structures may constitute memories of images, scenarios, music pieces, and spoken or written words, sentences, stories or books, to name a few implementations.

It might be noted that both the theory of prime numbers, intriguing human minds for centuries, and the theory of random graphs, presented in the mid 20th century, have revealed truly astonishing results, many of which have impacted both natural science and technological advancement. Yet, until recently, neither seems to have been directly associated with neuroscience research. More specifically, while the product-periods of primes have been noted in the contexts of music (du Sautoy, 2003), and meta-cyclic products of primes and non-primes have been noted in the natural survival game of predator and prey (Williams and Simon, 1995), such meta-periods have not been associated with cortical activity. Baram (2020a) has suggested a role for prime numbers in the categorization of neural circuit structures in accordance with their membrane and synapse polarities. Here, the meta-periodic effect of relatively small primal-size neural circuits interaction, generated by the combined inhibitory nature of circuit interneurons (Markram *et al.*, 2004) and excitatory nature of inter-circuit activity (Battaglia *et al.*, 2012) is noted. Further, random tree networks have been recently suggested as a model for neuronal connectivity, resulting in random "small world" structures (Ajazi *et al.*, 2019). We have shown that the meta-periodic firing-rate dynamics resulting from the cyclic connectivity of primal size circuits proposed in the present chapter constitutes a highly structured cortical language. Specifically, we have shown that the estimated number of neurons in the human brain limits the majority of neural circuits to primal size seven or less, with relatively little room to spare for circuits of larger primal sizes. As indicated by Table 6.2, this would allow, at the same time, for a minority of very long (up to several hundreds of millions of circuits) meta-periods and a majority of relatively short (up to several hundreds of circuits) meta-periods. The latter category includes single circuits of primal size seven or less, ratifying experimental findings on the human brain's capacity of working memory (Miller, 1956; Pribram *et al.*, 1960).

Part II
Firing-rate Linguistics

Chapter 7

Firing-rate Dynamics Irreversibility

7.1 Introduction

Irreversibility of natural processes is implied by the second law of thermodynamics (Mandl, 1988). In formal terms, the mathematical models of such processes are non-invertible. Here, we show that the widely recognized, experimentally supported model of neural firing-rate dynamics is non-invertible, hence, time-irreversible. This fundamental inability to reveal past values of a firing-rate process from present values would seem to present an inherent behavioral difficulty, as present behavior should somehow employ past experience. Luckily, as stressed in this book, memories of past experience are encapsulated in the form of global dynamic attractors, rather than actual values of firing-rates. The non-invertibility of the firing-rate process motivates, then, the central role played by the singularities of the model as a language of cortical dynamics. To this end, we make room for a detailed proof of the non-invertibility of the model, as revealed by an earlier work (Baram, 2012).

7.2 Discrete iteration map of neural firing and synaptic plasticity

The vector-valued truncated discrete-time version of the spike response model, Eq. (2.15), for the firing-rates is given by

$$v(k+1) = K_v v(k) + T_v^{-1} f\left(W(k)^T \sum_{p=0}^{N-1} E_p^{-1} v(k-p) + u(k) \right) \qquad (7.1)$$

where $v(k)$ is the vector of firing-rates corresponding to the neurons at time instance k, $W(k)$ is the matrix of synaptic weights, $u(k)$ is the vector of conductance-based activation inputs, $K_v = \text{diag}[(\tau_{v_i} - 1)/\tau_{v_i}]_{i=1...n}$, $T_v = \text{diag}[\tau_{v_i}]_{i=1...n}$ and $E_p = \text{diag}[e^{p/\tau_i}]_{i=1...n}$ are diagonal matrices of the corresponding elements, $f(x) = \text{vec}[f(x_i)]_{i=1...n}$ and $u(k) = \text{vec}[u_i(k)]_{i=1...n}$ are vectors of the corresponding elements, and $W(k) = [\omega_1(k) \ \omega_2(k) \ \cdots \ \omega_n(k)]$ is a matrix whose columns are $\omega_i(k)$, $i = 1 \ldots n$. N is a positive integer which satisfies the desired relative error ρ_i in Eq. (2.23) for any i, and f is a vector-valued function, whose scalar elements are defined by f_i of Eq. (2.16).

A discrete-time matrix-valued version of Eq. (2.17) is given by

$$W(k+1) = W(k) + T_\omega^{-1} \sum_{p=0}^{N-1} E_p^{-1}[[v(k-p) - \theta(k)]v^2(k-p)^T] * S$$

$$(7.2)$$

with

$$\theta(k) = T_\theta^{-1} \sum_{p=0}^{N-1} E_p^{-1} v^2(k-p) \qquad (7.3)$$

$T_\omega = \text{diag}[\tau_{\omega_i}]_{i=1...n}$, $T_\theta = \text{diag}[\tau_{\theta_i}]_{i=1...n}$ and $S = [s_1 \ s_2 \cdots s_n]$, and where $v^2(k-p)$ is the vector whose i'th element is $v_i^2(k-p)$ and where S is the connectivity matrix, having the value 1 in all positions corresponding to connected neurons and the value 0 elsewhere, and $*$ denotes the element-wise product between two matrices of the same dimension, as in

$$A * X = \begin{bmatrix} a_{1,1}x_{1,1} & a_{1,2}x_{1,2} & \cdots & a_{1,r}x_{1,r} \\ a_{2,1}x_{2,1} & a_{2,2}x_{2,2} & \cdots & a_{2,r}x_{2,r} \\ \vdots & & & \\ a_{p,1}x_{p,1} & a_{p,2}x_{p,2} & \cdots & a_{p,r}x_{p,r} \end{bmatrix} \qquad (7.4)$$

with $a_{i,j}$ and $x_{i,j}$ the respective scalar elements.

7.3   Non-invertibility (time-irreversibility) of the map

The map, Eqs. (7.1)–(7.3), is invertible if and only if it has an inverse which yields, for any values of the map parameters and any real values of the firing-rates at time $k + 1$, unique real values for the firing-rates at time k. If this is not the case, the map is said to be non-invertible. It should be noted that only positive (including zero) values are acceptable for the firing-rates.

Theorem 1. *The map, Eqs. (7.1)–(7.3), is non-invertible.*

Proof.   Denoting

$$v_p(k) = v(k - p) \tag{7.5}$$

Eq. (7.1) can be written as

$$v_0(k + 1) = K_v v_0(k) + T_v^{-1} f \left(W(k)^T \sum_{p=0}^{N-1} E_p^{-1} v_p(k) + u(k) \right) \tag{7.6}$$

and Eq. (7.2) can be written as

$$W(k + 1) = W(k) + T_\omega^{-1} \sum_{p=0}^{N-1} E_p^{-1} \left[v_p(k) v_p^2(k)^T \right] * S$$

$$- T_\omega^{-1} \sum_{p=0}^{N-1} E_p^{-1} \left[\theta(k) v_p^2(k)^T \right] * S \tag{7.7}$$

where, by Eq. (7.3),

$$\theta(k) = T_\theta^{-1} \sum_{p=0}^{N-1} E_p^{-1} v_p^2(k) \tag{7.8}$$

Substituting Eq. (7.7) and Eq. (7.8) into Eq. (7.6) and employing Eq. (7.5), we obtain

$$v_0(k+1) = K_v v_0(k) + T_v^{-1} f\left(P\left(v_p(k), p = 0, \ldots, N-1\right)\right)$$
$$v_1(k+1) = v_0(k)$$
$$v_2(k+1) = v_1(k)$$
$$\vdots$$
$$v_{N-1}(k+1) = v_{N-2}(k) \tag{7.9}$$

where

$$P\left(v_p(k), p = 0, \ldots, N-1\right)$$
$$= -\left[\sum_{p=0}^{N-1} S^T * \left[v_p^2(k) \sum_{p=0}^{N-1} v_p^2(k)^T E_p^{-1}\right] T_\theta^{-1} E_p^{-1}\right]$$
$$\times T_\omega^{-1} \sum_{p=0}^{N-1} E_p^{-1} v_p(k)$$
$$+ \left[\sum_{p=0}^{N-1} S^T * [v_p(k) v_p^2(k)^T] E_p^{-1}\right] T_\omega^{-1} \sum_{p=0}^{N-1} E_p^{-1} v_p(k)$$
$$+ W(k+1)^T \sum_{p=0}^{N-1} E_p^{-1} v_p(k) + u(k) \tag{7.10}$$

Given a unique positive real value for $v_i(k+1), i = 1, \ldots, N-1$, the last $N-1$ equations of Eq. (7.9) yield unique positive real values for $v_i(k), i = 0, \ldots, N-2$. It remains to determine whether, given in addition $v_0(k+1)$, Eq. (7.9) provides a unique positive real value for $v_{N-1}(k)$. The threshold non-linearity Eq. (2.16) and the first equation of Eq. (7.9) imply that either

$$P(v_p(k), p = 0, \ldots, N-1) < 0 \tag{7.11}$$

or

$$v_0(k+1) = K_v v_0(k) + T_v^{-1}(P(v_p(k), p = 0, \ldots, N-1)) \quad (7.12)$$

If Eq. (7.11) is satisfied, then, by the first equation of Eq. (7.9)

$$v_0(k+1) = K_v v_1(k+1) \quad (7.13)$$

Hence, Eq. (7.12) is satisfied in the range

$$v_0(k+1) \neq K_v v_1(k+1) \quad (7.14)$$

In order to show that the map is non-invertible, it will suffice to show that Eq. (7.12) does not have a unique real-valued solution for $v_{N-1}(k)$. Substituting Eq. (7.10) into Eq. (7.12), we have

$$T_v^{-1} \left[\sum_{p=0}^{N-1} S^T * \left[v_p^2(k) \sum_{p=0}^{N-1} v_p^2(k)^T E_p^{-1} \right] T_\theta^{-1} E_p^{-1} \right]$$

$$\times T_\omega^{-1} \sum_{p=0}^{N-1} E_p^{-1} v_p(k)$$

$$- T_v^{-1} \left[\sum_{p=0}^{N-1} S^T * [v_p(k) v_p^2(k)^T] E_p^{-1} \right] T_\omega^{-1} \sum_{p=0}^{N-1} E_p^{-1} v_p(k)$$

$$- K_v v_0(k) - T_v^{-1} W(k+1)^T \sum_{p=0}^{N-1} E_p^{-1} v_p(k)$$

$$- T_v^{-1} \beta(k) - v_0(k+1) = 0 \quad (7.15)$$

The left-hand side of Eq. (7.15) can be written as a polynomial in $v_{N-1}(k)$

$$P(v_{N-1}(k)) = P_5(v_{N-1}(k)) + P_4(v_{N-1}(k)) + P_4(v_{N-1}(k))$$

$$+ P_3(v_{N-1}(k)) + P_1(v_{N-1}(k)) + P_0(v_{N-1}(k)) \quad (7.16)$$

where, incorporating Eq. (7.9),

$$P_5(v_{N-1}(k)) = T_v^{-1} \left[S^T * [v_{N-1}^2(k)v_{N-1}^2(k)^T] \right]$$
$$\times E_{N-1}^{-1} T_\theta^{-1} E_{N-1}^{-1} T_\omega^{-1} E_{N-1}^{-1} v_{N-1}(k)$$

$$P_4(v_{N-1}(k)) = T_v^{-1} \left[S^T * [v_{N-1}^2(k)v_{N-1}^2(k)^T] \right]$$
$$\times E_{N-1}^{-1} T_\theta^{-1} E_{N-1}^{-1} T_\omega^{-1} \sum_{p=1}^{N-1} E_p^{-1} v_p(k+1)$$

$$P_3(v_{N-1}(k)) = - T_v^{-1} \left[S^T * [v_{N-1}^2(k)v_{N-1}^2(k)^T] \right]$$
$$\times E_{N-1}^{-1} T_\omega^{-1} E_{N-1}^{-1} v_{N-1}(k)$$

$$P_2(v_{N-1}(k)) = - T_v^{-1} \left[S^T * [v_{N-1}^2(k)v_{N-1}^2(k)^T] \right]$$
$$\times E_{N-1}^{-1} T_\omega^{-1} E_{N-1}^{-1} \sum_{p=1}^{N-1} E_p^{-1} v_p(k+1)$$

$$P_1(v_{N-1}(k)) = T_v^{-1} \omega(k+1)^T E_{N-1}^{-1} v_{N-1}(k)$$

$$P_0(v_{N-1}(k)) = T_v^{-1} \left[\sum_{p=1}^{N-1} S^T * \left[v_{N-1}^2(k+1) \sum_{p=1}^{N-1} v_p^2(k+1)^T E_p^{-1} \right. \right.$$
$$\left. \left. \times T_\theta^{-1} E_p^{-1} \right] T_\omega^{-1} \sum_{p=1}^{N-1} E_p^{-1} v_p(k+1) \right.$$

$$- T_v^{-1} \left[\sum_{p=1}^{N-1} S^T * [v_p(k+1)v_p^2(k+1)^T] E_p^{-1} \right]$$
$$\times T_\omega^{-1} \sum_{p=1}^{N-1} E_p^{-1} v_p(k+1)$$

$$+ T_v^{-1} \left[W(k+1)^T \sum_{p=1}^{N-1} E_p^{-1} v_p(k+1) \right]$$

$$+ K_v v_1(k+1) - v_0(k+1) + T_v^{-1} u(k) \tag{7.17}$$

The i'th row of the vector Eq. (7.15) is a scalar polynomial equation of order 5 in the elements of $v_{i,N-1}(k)$. The terms of Eq. (7.17) corresponding to the scalar polynomial are, in descending order of power of $v_{i,N-1}(k)$,

$$p_{i,5}(v_{N-1}(k)) = \frac{e^{-3(N-1)/\tau_i}}{\tau_{v_i}\tau_{\theta_i}\tau_{\omega_i}} v_{i,N-1}^2(k) \sum_{j=1}^{n} v_{j,N-1}^3(k)$$

$$p_{i,4}(v_{N-1}(k)) = \frac{e^{-2(N-1)/\tau_i}}{\tau_{v_i}\tau_{\theta_i}\tau_{\omega_i}} v_{i,N-1}^2(k)$$

$$\times \sum_{j=1}^{n} \left[v_{j,N-1}^2(k) \sum_{p=0}^{N-2} e^{-p/\tau_j} v_{j,p}(k) \right]$$

$$p_{i,3}(v_{N-1}(k)) = -\frac{e^{-2(N-1)/\tau_i}}{\tau_{v_i}\tau_{\omega_i}} v_{i,N-1}(k) \sum_{j=1}^{n} v_{j,N-1}^3(k)$$

$$p_{i,2}(v_{N-1}(k)) = -\frac{e^{-2(N-1)/\tau_i}}{\tau_{v_i}\tau_{\omega_i}} v_{i,N-1}(k)$$

$$\times \sum_{j=1}^{n} \left[v_{j,N-1}^2(k) \sum_{p=0}^{N-2} e^{-p/\tau_j} v_{j,p}(k) \right]$$

$$p_{i,1}(v_{N-1}(k)) = \frac{e^{-(N-1)/\tau_i}}{\tau_{v_i}} \sum_{j=1}^{n} \omega_{i,j}(k+1)v_{j,N-1}(k)$$

$$p_{i,0}(v_{N-1}(k)) = [\mu(k+1)]_i + \left[T_v^{-1} u(k) \right]_i \tag{7.18}$$

where $\omega_{i,j}(k+1)$ is the i,j'th scalar element of $W(k+1)$, and $\left[T_v^{-1}u(k) \right]_i$ and $[\mu(k+1)]_i$ are the $i'th$ scalar elements of the vectors $T_v^{-1}u(k)$ and

$$\mu(v(k+1)) = T_v^{-1} \left[\sum_{p=1}^{N-1} S^T * \left[v_p^2(k+1) \sum_{p=1}^{N-1} v_p^2(k+1)^T E_p^{-1} \right. \right.$$

$$\left. \left. \times T_\theta^{-1} E_p^{-1} \right] T_\omega^{-1} \sum_{p=1}^{N-1} E_p^{-1} v_p(k+1) \right.$$

$$-T_v^{-1}\left[\sum_{p=1}^{N-1} S^T * [v_p(k+1)v_p^2(k+1)^T] E_p^{-1}\right]$$

$$\times T_\omega^{-1} \sum_{p=1}^{N-1} E_p^{-1} v_p(k+1)$$

$$-T_v^{-1}\left[W(k+1)^T \sum_{p=1}^{N-1} E_p^{-1} v_p(k+1)\right]$$

$$-K_v v_1(k+1) = v_0(k+1) \tag{7.19}$$

respectively. While the map at hand is polynomial in the firing-rates, it is also a transcendental (non-algebraic) function of the initial conditions and the parameter values (including, in our case, external activation inputs. Indeed, local bifurcations are caused by changes in these parameters). Reduction to circuits of identical neurons is justified by the Hebbian paradigm, which implies segregation of mutually asynchronous circuits or, as considered in a subsequent chapter, such segregation enforced by synapse polarization.

If all the neurons have the same parameter values and the same initial conditions, they must behave in the same manner. This means that within this parametric domain (the domain of symmetry) all the scalar rows of Eq. (7.15) represent the same scalar polynomial equation. If the $i'th$ neuron is self-connected, the i'th row of Eq. (7.15) can be written as a univariate polynomial equation in $v_{i,N-1}(k)$ whose terms are, in descending order of power,

$$p_{i,5}(v_{i,N-1}(k)) = \frac{ne^{-3(N-1)/\tau_i}}{\tau_{v_i}\tau_{\theta_i}\tau_{\omega_i}} v_{i,N-1}^5(k)$$

$$p_{i,4}(v_{i,N-1}(k)) = \frac{ne^{-2(N-1)/\tau_i}\sum_{p=0}^{N-2} e^{-p/r_j} v_{i,p}(k)}{\tau_{v_i}\tau_{\theta_i}\tau_{\omega_i}} v_{i,N-1}^4(k)$$

$$p_{i,3}(v_{i,N-1}(k)) = -\frac{ne^{-2(N-1)/\tau_i}}{\tau_{v_i}\tau_{\omega_i}} v_{i,N-1}^3(k)$$

$$p_{i,2}(v_{i,N-1}(k)) = -\frac{ne^{-2(N-1)/\tau_i} \sum_{p=0}^{N-2} e^{-p/\tau_j} v_{i,p}(k)}{\tau_{v_i} \tau_{\omega_i}} v_{i,N-1}^2(k)$$

$$p_{i,1}(v_{i,N-1}(k)) = \frac{ne^{-(N-1)/\tau_i} w(k+1)}{\tau_{v_i}} v_{i,N-1}(k)$$

$$p_{i,0}(v_{i,N-1}(k)) = \left[\mu(v(k+1)) + T_v^{-1} u(k)\right]_i \qquad (7.20)$$

(note that, due to the assumed parametric identity, all the synaptic weights are the same, hence, $w(k+1) = \omega_{i,j}(k+1); i,j = 1,\ldots,n$). If the neurons are self-disconnected, n will be replaced by $n-1$ in all but the term for $p_{i,1}(v_{N-1}(k))$. Descartes' rule of signs (Fine and Rosenberger, 1997) states that the number of real positive roots of a univariate polynomial equation with real coefficients equals the number of sign changes, or less than it by a multiple of two, when the coefficients are ordered according to descending powers of the polynomial. It can be seen that, since $v_{i,N-1}(k)$ is positive, there is an even number of Descartes sign changes in Eq. (7.20) within the domain

$$\left[\mu(v(k+1)) + T_v^{-1} u(k)\right]_i > 0 \qquad (7.21)$$

In order to see that this is true, let us write Eq. (7.16) for the scalar case as

$$p(v_{N-1}(k)) = p_{i,5}(v_{N-1}(k)) + p_{i,4}(v_{N-1}(k)) + p_{i,3}(v_{N-1}(k))$$
$$+ p_{i,2}(v_{N-1}(k)) + p_{i,1}(v_{N-1}(k)) + p_{i,0}(v_{N-1}(k))$$
$$(7.22)$$

which can be rearranged in the order of decreasing powers of $v_{N-1}(k)$ as

$$p(v_{N-1}(k)) = p_{i,5}(v_{N-1}(k)) + [p_{i,4}(v_{N-1}(k))$$
$$+ p_{i,3}(v_{N-1}(k))] + p_{i,2}(v_{N-1}(k))$$
$$+ p_{i,1}(v_{N-1}(k)) + p_{i,0}(v_{N-1}(k)) \qquad (7.23)$$

It can be seen that, according to Eq. (7.20), the right-hand side of Eq. (7.23) has two Descartes' sign changes regardless of the sign of $[p_{i,4}(v_{N-1}(k)) + p_{i,3}(v_{N-1}(k))]$.

It follows that, within the domain Eq. (7.21), the map under parametric identity has no unique positive real inverse and hence is non-invertible. This holds for networks of any number of self-connected neurons, and for networks of two or more self-disconnected neurons (for a single self-disconnected neuron, all the terms of Eq. (7.20) but the last two vanish, and the map becomes invertible). □

7.4 Discussion

We have analytically shown that, depending on parameter values, the neural firing-rate map is non-invertible. Non-invertibility of the map motivates the global attractor approach taken in this book for explaining and demonstrating cortical dynamics in memory and other cortical functions, such as sensorimotor control. As the mathematical object under consideration — namely, the neural firing-rate model — is quite specific, we were able to avoid general, and rather abstract, concepts associated with the theory of non-invertible maps (e.g., Mira, 2007), and address the problem at hand in explicit terms. Furthermore, as shown in subsequent chapters of this book, reduction to circuits of identical neurons is justified by segregation of mutually asynchronous circuits into internally synchronous circuits, either by the Hebbian paradigm, or by synapse polarization. This reduction has, in turn, facilitated direct application of a fundamental rule, namely, Descartes' rule of signs, which applies to scalar-valued polynomials. It might be noted in this context that a certain extension of Descartes' rule of signs to the multivariate case has been suggested (Bihan and Dickenstein, 2017), but does not appear to have a formal journal citation.

Chapter 8

Discrete Iteration Maps of Neural Firing in Cortical Development

8.1 Introduction

Neurophysiological and molecular studies have distinguished between pre-critical period excitability, instrumental in initial circuit formation (Hsia *et al.*, 1998; Hensch *et al.*, 1998; Hensch, 2005; Tessier and Broadie, 2009; Ashby and Isaac, 2011; Gibson and Ma, 2011; Weiner *et al.*, 2013) and persistent plasticity, evidenced during critical development of ocular dominance in early life (Hensch, 2005; Hooks and Chen, 2007; Miyata *et al.*, 2012). Molecular mechanisms for activity-independent cortical plasticity control have been suggested, with both excitatory and inhibitory synapses playing critical roles (Constantine-Paton *et al.*, 1990; Huang *et al.*, 1999; Fagiolini and Hensch, 2000; Goold and Nicoll, 2010; Phillips *et al.*, 2011; Miyata *et al.*, 2012; Wang *et al.*, 2012). On the other hand, activity-dependent inter-neural connectivity has been supported by the Hebbian paradigm (Caporale and Yang, 2008; Buzsáki, 2010; Doll and Broadie, 2014).

It is largely believed that initial excitability enhances sensitivity to inhibitory effects, holding the key to plasticity modification (Huang *et al.*, 1999; Fagiolini and Hensch, 2000; Feller and Scanziani, 2005; Goold and Nicoll, 2010; Wang *et al.*, 2012). While pre-critical plasticity is marked by excitatory initiation, persistent plasticity in critical period is initiated when one inhibition level is crossed, and terminates, allowing for convergent plasticity to settle in, when a second inhibition level is crossed (Huang *et al.*, 1999; Fagiolini and Hensch, 2000; Miyata *et al.*, 2012). At the offset of persistent plasticity, triggered by experience, there is a sharp switch in the sign of plasticity (Wang *et al.*, 2012). The conditions for persistent plasticity, for convergent plasticity, and for no plasticity at all are met when excitatory and inhibitory inputs reach a corresponding balance (Fagiolini and Hensch, 2000; Maffei and Turrigiano, 2008). Sensory inputs have been shown to evoke on-going shunting in visual cortex circuits (Borg-Graham *et al.*, 1998). Mathematically, early studies of neural network dynamics have employed continuous time models (Wilson and Cowan, 1972; Cohen and Grossberg, 1983; Peterfreund and Baram, 1998a,b). The combined effects of firing-rate and plasticity time constants on firing-rate dynamics corresponding to different developmental stages have been analyzed, laying the ground for analytic unification with respect to neuronal properties, on the one hand, and cortical development, on the other (Baram, 2017a). As previous studies of cortical development have been essentially experimental, a formal distinction between excitability and persistence appears to have been rather evasive. In this, and the following two chapters, we relate the notions of excitatory, inhibitory, persistent and convergent plasticity to mathematical attributes of the underlying dynamical model.

8.2　Reduced discrete iteration maps

It has been noted that the mathematically imposing integral term in Eq. (2.15) is biologically and functionally insignificant (Wilson and Cowan, 1972), and that the essence of neural firing dynamics is captured by its instantaneous manifestation (Dayan and Abbott, 2001; Miller and Fumarola, 2012). Experimental imaging has shown that neurons having the same response types develop increased local connectivity, while neurons of different response types show no such connectivity (Kenet *et al.*, 2003; Karlsson and Frank, 2009; Komiyama *et al.*, 2010; Garner and Mayford, 2012). Furthermore, synchronous firing, which may conceptually arise under uniform parameterization and activation across circuit neurons, has been empirically observed in small (e.g., Marder and Bucher, 2001) and large (e.g., Lopes da Silva, 1991) circuits. As the presence of a general time-varying $u_i(t)$ does not seem to allow the revelation of the fundamental dynamic modes of firing-rate, we replace it for the purposes of this study by a constant value, u_i, as previously done in studies of firing-rate dynamics (e.g., Jolivet *et al.*, 2004). For circuits or subcircuits of n neurons firing in synchrony (with other possible circuit neurons filtered out), and constant activation u, Eq. (2.15) reduces the instantaneous scalar model

$$\tau_{m_i} \frac{d}{dt} v_i(t) = -v_i(t) + f_i(n\omega_i(t)v_i(t) + u_i) \tag{8.1}$$

where n is the number of neurons in the circuit that fire synchronously with the i'th neuron, $\omega_i(t)$ is the synaptic weight connecting each of these neurons to the i'th neuron (which, by the Hebbian filtering property noted above, is the same for all synapses connecting neurons interacting with the i'th neuron in a feedback fashion, including a possible self-connection). It might also be noted that while, as argued above, the model Eq. (8.1) arises generally in the context of synchronously interacting neurons, it may also be directly associated, for $n = 1$, with certain neuron types involving self-feedback, specifically, the autapse (Van der Loos and Glaser, 1972).

The BCM plasticity rule (Bienenstock *et al.*, 1982), enhanced by stabilizing modifications (Intrator and Cooper, 1992; Cooper *et al.*, 2004), is a widely recognized, biologically plausible, mathematical representation of the Hebbian learning paradigm (Hebb, 1949). It suggests the computation of $\omega_i(t)$ by the model

$$\tau_{\omega_i} \frac{d}{dt} \omega_i(t) = -\omega_i(t) + (v_i(t) - \theta_i(t))v^2(t) \tag{8.2}$$

where τ_{ω_i} is the learning time constant and $\theta_i(t)$ is a variable threshold satisfying (Intrator and Cooper, 1992; Cooper *et al.*, 2004)

$$\theta_i(t) = \frac{1}{\tau_{\theta_i}} \int_{-\infty}^{t} v^2(\tau) \exp(-(t-\tau)/\tau_{\theta_i})d\tau \tag{8.3}$$

where τ_{θ_i} is the thresholding time constant.

The discretization of a continuous-time dynamical system (e.g., Qwakenaak and Sivan, 1972) is done under the assumption that the time steps are sufficiently small so as to allow the approximation of the input by a constant value throughout the time step. Such approximation may, in principle, introduce inaccuracies in the description of the system's behavior. Yet, discretization is often used as computational means as long as it does not introduce instability. In the present context of neural behavior, the validity of the discrete time model can be qualitatively tested by comparison of simulations to the ample evidence produced by neurobiological experiments. The discrete time versions under unity time steps of Eqs. (8.1)–(8.3), omitting the index i for notational convenience, are

$$v(k) = \alpha v(k-1) + \beta f(n\omega(k)v(k-1) + u) \tag{8.4}$$

with

$$\alpha = \exp(-1/\tau_m), \ \beta = 1 - \alpha, \tag{8.5}$$

$$\omega(k) = \varepsilon\omega(k-1) + \gamma(v(k-1) - \theta(k-1))v^2(k-1) \tag{8.6}$$

and

$$\theta(k) = \delta \sum_{i=0}^{N} \exp(-i/\tau_\theta)v^2(k-i) \tag{8.7}$$

with

$$\varepsilon = \exp(-1/\tau_\omega), \gamma = 1 - \varepsilon, \delta = 1/\tau_\theta \tag{8.8}$$

8.3   Discrete iteration maps in developmental stages

The discrete-time firing-rate models are now specified for each of the developmental stages, corresponding to hypothesized domains of the time constants involved.

8.3.1   *Pre-critical period excitability*

Hypothesizing fast cortical response in early life (as suggested by numerous publications in the popular literature with, apparently, no specific neurophysiological support), we assume near-zero values of the time constants τ_m, τ_w and τ_θ. Letting $\tau_m \to 0$, Eq. (8.4) becomes

$$v(k) = f(nw(k)v(k-1) + u) \tag{8.9}$$

Letting $\tau_w \to 0$ and $\tau_\theta \to 0$, we have an "instantaneously sliding threshold", yielding the continuous-time plasticity model $dw(t)/dt = (1 - v(t))v(t)^3$ (Cooper *et al.*, 2004, p. 23) and, under Eqs. (8.6) and (8.7), its discrete-time version (Qwakenaak and Sivan, 1972, pp. 445–447)

$$w(k) = (1 - v(k-1))v^3(k-1) \tag{8.10}$$

A combination of Eqs. (8.9) and (8.10) yields

$$v(k) = \begin{cases} f_1(v(k-1)) = 0 & \text{for } m(v(k-1)) \le 0 \\ f_2(v(k-1)) = m(v(k-1)) & \text{for } m(v(k-1)) > 0 \end{cases} \tag{8.11}$$

where

$$m(v(k-1)) = -nv^5(k-1) + nv^4(k-1) + u \tag{8.12}$$

A positive slope of the map represents excitatory synaptic plasticity, while a negative slope represents inhibitory plasticity, as does a zero value. The map Eq. (8.11) has a single maximum, h, at

$$v(k-1) = \frac{4}{5} \tag{8.13}$$

yielding

$$h = +\left[\left(\frac{4}{5}\right)^4 - \left(\frac{4}{5}\right)^5\right]n + u \tag{8.14}$$

which defines its inhibitory onset. The inhibitory offset g is the threshold which, by Eq. (8.11), satisfies $m(v(k-1)) = 0$, yielding, by Eq. (8.12), the polynomial equation

$$ng^5 - ng^4 - u = 0 \qquad (8.15)$$

The map has a single fixed point, for which $v(k) = v(k-1)$, at p, satisfying

$$np^5 - np^4 + p - u = 0 \qquad (8.16)$$

The slope of the map at the fixed point p is

$$\lambda(p) = -5np^4 + 4np^3 \qquad (8.17)$$

The map Eq. (8.11), representing the neuronal firing-rate under pre-critical plasticity, is depicted in Fig. 8.1(a) for the parameters $n = 2, u = 1, \tau_m = 0, \tau_w = 0, \tau_\theta = 0$. It can be seen that the pre-critical setting allows for excitatory plasticity for low firing-rate values, turning inhibitory at h.

8.3.2 *Critical period persistent plasticity*

As it is often assumed that changes in synaptic properties are markedly faster than changes in membrane properties (e.g., $\tau_m \gg \tau_w$, Brunel and Sergi, 1998; Dayan and Abbott, 2001; Miller and Fumarola, 2012), the transition from pre-critical to critical period (Huang *et al.*, 1999; Fagiolini and Hensch, 2000; Miyata *et al.*, 2012) can be represented by an increase in τ_m, while τ_w and τ_θ maintain near-zero values. This implies that $v(k)$ becomes governed by Eq. (8.4), while $w(k)$ remains governed by Eq. (8.10), which implies

$$v(k) = \begin{cases} f_1\left(v(k-1)\right) & \text{for } \beta m(v(k-1)) \leq 0 \\ f_2\left(v(k-1)\right) & \text{for } \beta m(v(k-1)) > 0 \end{cases} \qquad (8.18)$$

where

$$f_1\left(v(k-1)\right) = \alpha v(k-1) \qquad (8.19)$$

and

$$f_2\left(v(k-1)\right) = \beta m(v(k-1)) + \alpha v(k-1) \qquad (8.20)$$

with $m(v(k-1))$ satisfying Eq. (8.12). The map represented by Eq. (8.18) has a single maximum, h, at the point $v(k-1)$ satisfying

$$5\beta v(k-1)^4 - 4\beta v(k-1)^3 - \alpha = 0 \tag{8.21}$$

The bend point g in the map Eq. (8.18) satisfies the equation $g = m(v(k-1)) = 0$, while the single fixed point, p satisfies the equation $v(k-1 = \beta m(v(k-1)) + \alpha v(k-1)$. As, by Eq. (8.10), $\omega(k)$ is an undamped polynomial function of $v(k-1)$, the dynamic modes of $\omega(k)$ can be expected to have an inherent similarity to those of $v(k)$, and the synaptic plasticity under consideration will persist as long as the neuronal firing does. From a functional viewpoint, high plasticity is required for circuit formation and fast learning in early development. The bi-functional discrete iteration map Eq. (8.18) represents the neuronal firing-rate under persistent plasticity, depicted in Fig. 8.1(b) for the parameters $n = 2$, $u = 1$, $\tau_m = 1$, $\tau_\omega = 0$, $\tau_\theta = 0$. As will be shown in subsequent chapters, the dynamic nature of neural firing is highly dependent on the interplay between the excitatory (positive slope) and the inhibitory (negative slope) parts of the map. As shown in the figure, the transition points between these parts are h and g, representing a local maximum and a local minimum of the map, respectively.

8.3.3 *Convergent plasticity and synaptic maturity*

For a non-zero synaptic time constant τ_ω, variants of the Hebbian learning paradigm (Hebb, 1949), represented here by the BCMI (Bienenstock *et al.*, 1982; Intrator and Cooper, 1992) Eqs. (8.2) and (8.3), have been shown to converge to constant synaptic weights (Rosenblatt, 1958; Oja, 1982; Cooper *et al.*, 2004). Synaptic convergence has been identified as a developmental stage immediately following critical development and persisting into synaptic maturity (Constantine-Paton *et al.*, 1990; Phillips *et al.*, 2011). Governed by the model Eqs. (8.4), (8.6) and (8.7), such convergence yields a limit value ω for the synaptic weight, which will not be changed by further activity. Replacing $\omega(k)$ by ω in the convergent plasticity version of the model Eqs. (8.4) and (8.6), the firing-rate model under synaptic maturity attains the bi-linear form

$$v(k) = \begin{cases} f_1\left(v(k-1)\right) = \lambda_1 v(k-1) & \text{for } \beta n \omega v(k-1) + \beta u \le 0 \\ f_2\left(v(k-1)\right) = \lambda_2 v(k-1) + \beta u & \text{for } \beta n \omega v(k-1) + \beta u > 0 \end{cases} \tag{8.22}$$

with

$$\lambda_1 = \alpha \tag{8.23}$$

and

$$\lambda_2 = \alpha + \beta n \omega \tag{8.24}$$

The inhibitory domain is bounded by

$$h = \beta u \tag{8.25}$$

on the left and by

$$g = \frac{\beta u}{\lambda_1 - \lambda_2} \tag{8.26}$$

on the right, and the fixed point is at

$$p = \frac{\beta u}{1 - \lambda_2} \tag{8.27}$$

The bi-linear map, Eq. (8.22), representing synaptic maturity, is depicted in Fig. 8.1(c) for the parameters $n = 2$, $u = 1.2$, $\tau_m = 3$, $\tau_\omega = 0.5$, $\tau_\theta = 0.01$ and $\omega = -0.8321$.

8.3.4 *Synaptic rigidity*

As noted above, the state of invariant synaptic weight can be reached by synaptic convergence. It can also result, however, from a very large synaptic time constant τ_ω, representing a very slow response, which, associated with old age, may be termed synaptic rigidity (McGahon *et al.*, 1999). This will yield, by Eq. (8.8), $\varepsilon = 1$, $\gamma = 0$, hence,

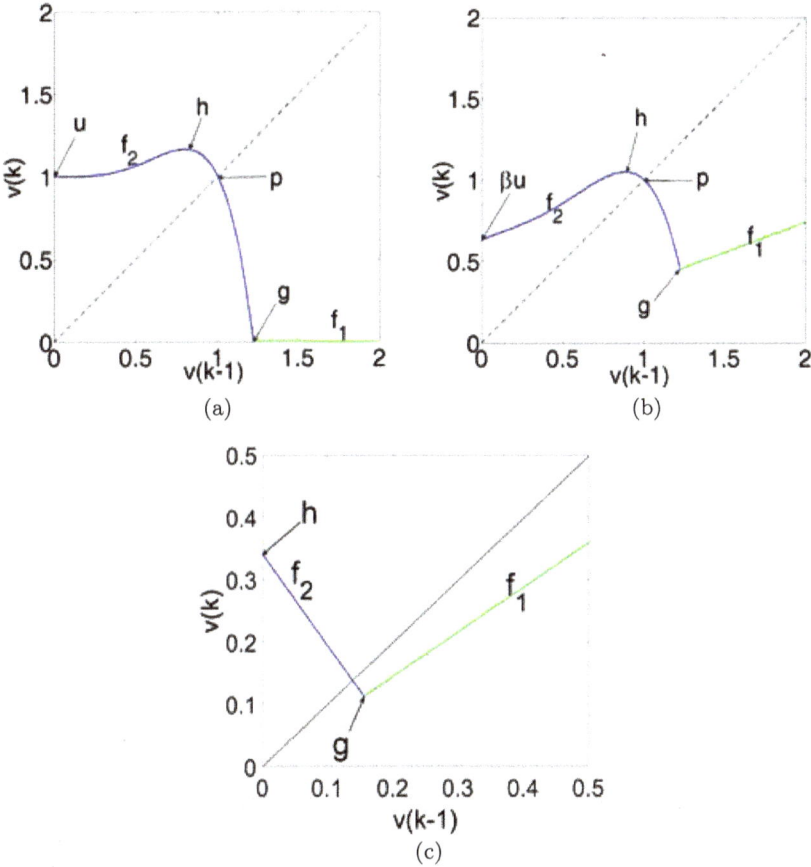

Figure 8.1. Developmental plasticity maps of neuronal firing: (a) pre-critical plasticity, (b) critical persistent plasticity, (c) synaptic maturity and rigidity. The parts f_1 and f_2 of the bi-functional map f are defined in the text for each of the developmental stages (Eq. (8.11) for the pre-critical stage, Eq. (8.18) for the critical stage, and Eq. (8.22) for the convergence and the rigidity stages, all with the specified parameter values). The inhibition onset and termination thresholds are h and g, where the slope of the map changes sign. The transition from persistent plasticity to synaptic maturity is mediated by convergent plasticity, pushing h to the left, where it meets the intersection point βu of f_2 with the axis $v(k)$. The function f_1 is governed by Eq. (8.11) in case (a), and by Eq. (8.19) in the other cases (the value of ω was obtained by running the model Eqs. (8.4)–(8.8) to convergence, achieved by $N = 100$).

by Eq. (8.6),

$$\omega(k) = \omega(k-1) = \omega \tag{8.28}$$

While the firing-rate model Eq. (8.22) applies to both synaptic maturity and synaptic rigidity, both represented by Fig. 8.1(c), it should be noted that while synaptic convergence leading to synaptic maturity facilitates synaptic learning, forming the neuronal firing code, synaptic rigidity, which may arise prematurely due to pathology, blocks out synaptic learning. This does not eliminate the possibility of learning by circuit modification, detailed in a later chapter. As in the rest of this book the analysis and results corresponding to synaptic maturity will, in essence, apply equally to synaptic rigidity, we will avoid a separate mention of synaptic rigidity and refer to synaptic maturity for both developmental stages.

8.4   Discussion

Coupling synaptic plasticity and firing-rate models reveals a more comprehensive framework of cortical plasticity than that widely associated with synaptic weights. We have shown that the discrete iteration maps of synaptically-modulated neural firing-rate depend on the corresponding synapse and membrane time constants, which are hypothesized to increase with age. The developmental dependence of the map suggests a broader notion of "metaplasticity" (or "plasticity of plasticity"), which has been previously associated with activity-related changes in synaptic plasticity alone (Abraham and Bear, 1996; Abraham, 2008; Ming-Chia *et al.*, 2010). While the differences between the maps corresponding to different developmental stages are quite noticeable, asserting a large variety of experimental findings, the maps also reveal certain common age-independent attributes. Indeed, while certain cortical activities are more vigorously executed at early development, they can be clearly traced, at different levels, throughout life. While the relevance of some cortical functions increases with age, that of others decays. Changes in parameter values, specifically, time constants, activation levels and synaptic weights change the nature of a map even within developmental stages. Each map represents a mathematical subscription for dynamic modes of neural firing-rate. The correspondence between maps and their characteristic firing-rate modes is elaborated on in the next two chapters.

The analytic findings link the model and its parameters to known dynamic attributes associated with developmental stages. The discrete iteration maps corresponding to the developmental stages, analytically derived from the underlying models, closely (albeit qualitatively) adhere to transition points between and within these stages, as found in neurobiological experiments. Specifically, pre-critical plasticity is marked by excitatory initiation, represented in Fig. 8.1(a) by a positive slope of the map, while persistent plasticity in critical period is initiated when one inhibition level (represented by point h in Fig. 8.1(b)) is crossed, and terminates, allowing for convergent plasticity to settle in, when a second inhibition level (represented by point h in Fig. 8.1(c)) is crossed. These analytic observations are in agreement with experimental findings (Huang *et al.*, 1999; Fagiolini and Hensch, 2000; Miyata *et al.*, 2012). At the offset of persistent plasticity, triggered by experience, there is a sharp switch (represented by point g in Fig. 8.1(c)) in the sign of plasticity, also in agreement with experimental findings (Wang *et al.*, 2012). The conditions for persistent and convergent plasticity are met when excitatory and inhibitory inputs (hence, corresponding synapses) reach a certain balance (represented by point g in Figs. 8.1(b) and 8.1(c)), as experimentally suggested (Fagiolini and Hensch, 2000; Maffei and Turrigiano, 2008).

The correspondence between the dynamic attributes associated with developmental stages and the time constants of the underlying mathematical models of firing and plasticity is particularly striking. High sensitivity, associated with pre-critical and critical periods, is represented by near-zero τ_ω, τ_θ and τ_m values in pre-critical period, indicating very fast synapse and membrane responses, and by near-zero τ_ω and τ_θ values, and slightly higher τ_m value in critical period, indicating very fast synaptic response, and slightly slower membrane response. As can be seen in Fig. 8.1, the plasticity maps (a) and (b), corresponding to these early periods, both have an excitatory component (represented by positive slope of the map) corresponding to low values of $v(k-1)$ (values lower than the ones corresponding to point h), which, indicating high sensitivity to weak signals, stands in sharp contrast to the map 8.1(c), corresponding to the maturity and rigidity stages, having no such plasticity component. As can be seen in Fig. 8.1, the negative slopes in the said domains increase with h, hence, by Eq. (8.25), with u, which may be induced by external sensory signals, ratifying high sensitivity to sensory experience in early development (Hooks and Chen, 2007).

The developmental plasticity maps represented in Fig. 8.1 show clearly noticeable differences, yet certain similarities between their inherent characteristics, corresponding to the different developmental stages. It has been previously stated that, "the use of common vocabulary reflects the hope of neurobiologists that the developmental mechanisms of plasticity will prove similar to those underlying learning and memory in the mature brain" (Constantine-Paton *et al.*, 1990). A complete immersion of the theory in a biological context would require establishing firm links between the time constants and the molecular processes involved. Such linkage might be mediated by matching the firing modes observed in specifically designed biological experiments, subject to molecular and functional modifications, to those predicted by the analysis.

Chapter 9

Global Attractors of Neural Firing-rate in Early Development

9.1 Introduction

The dynamic nature of an attractor induced by a fixed point p, marking the intersection of the map with the diagonal $v(k) = v(k-1)$, is determined by λ_p, the slope of the map at p. For a scalar discrete iteration map , the dynamic nature of an attractor is revealed by the corresponding Koenigs-Lemeray cobweb diagram (Koenigs, 1884; Lemeray,1895; Knoebel, 1981; Abraham *et al.*, 1997), constructed as described in Chapter 2. As cobweb diagrams are restricted to scalar dynamics, we employ, for graphical illustration purposes, circuits of identical neurons, which, isolated by the Hebbian paradigm, collapse in firing dynamics to the scalar case, corresponding to synchronous circuit firing. The diversion from synchrony is discussed in Chapter 8. As, from a system-theoretic viewpoint, it is the system's response to constant external inputs, which characterizes the dynamic modes, we scan a wide range of u values (taken first to be between -10 and 10, then slowly increased to see if any other attractor types surface) so as to find a set of constant values that spell out the global attractors of neural firing-rate, at each developmental stage. The analytic results presented were ratified by both numerous cobweb diagrams and numerous simulations (only few of which are displayed) to be highly robust against local changes in the time constants and the activation values.

While the effects of synapse and membrane time constants on neuronal functions has been recognized (Connor and Stevens, 1971; Elson *et al.*, 2002), the possible correspondence between developmental stages and such time constants appears to have been explicitly addressed, or even noted, only recently (Baram, 2017a). As there do not appear to be any available empirical data on such time constants, or even on developmentally related time scales, our analysis, employing unity sampling intervals, may be considered qualitative in this respect. Yet, the emerging codes of global attractors are found to be consistent with numerous empirically observed firing modes of biological neurons.

9.2   Global attractors of neural firing-rate under pre-critical plasticity

The neural firing map, Eq. (8.11), corresponding to pre-critical plasticity, is characterized by near-zero time constants. The nature of its global attractor depends, essentially, on the activation level u. For a positive value of u, the slope of the map λ_p at the fixed point p, inducing the global attractor, will be negative. For a negative value of u, the global attractor will be at the origin, manifesting a silent neuron. Other global attractor types arise when u is positive. The corresponding global attractor types are specified below for $\tau_m = \tau_\omega = \tau_\theta = 0, n = 2$, and representative values of u, illustrated in Fig. 9.1.

(a) *Fixed-point attractor.* For $u = 0.8$, any initial condition, $v(0)$, will result in convergence of the corresponding cobweb trajectory to the fixed point p, as illustrated by Fig. 9.1(a).

(b) *Largely-cyclic attractor.* For $u = 1$, any initial condition, $v(0)$, will result in convergence of the cobweb trajectory to the semi-open intervals $[ap)$ and $(pb]$, (with the end points a and b defined by a three-step cobweb initiated at h) separated by the repelling point p, as illustrated in Fig. 9.1(b) by the red and violet intervals on the diagonal. We call this attractor type "largely-cyclic" since the firing-rate process will sequentially alternate between the two indicated intervals, possibly performing a precise oscillation, or

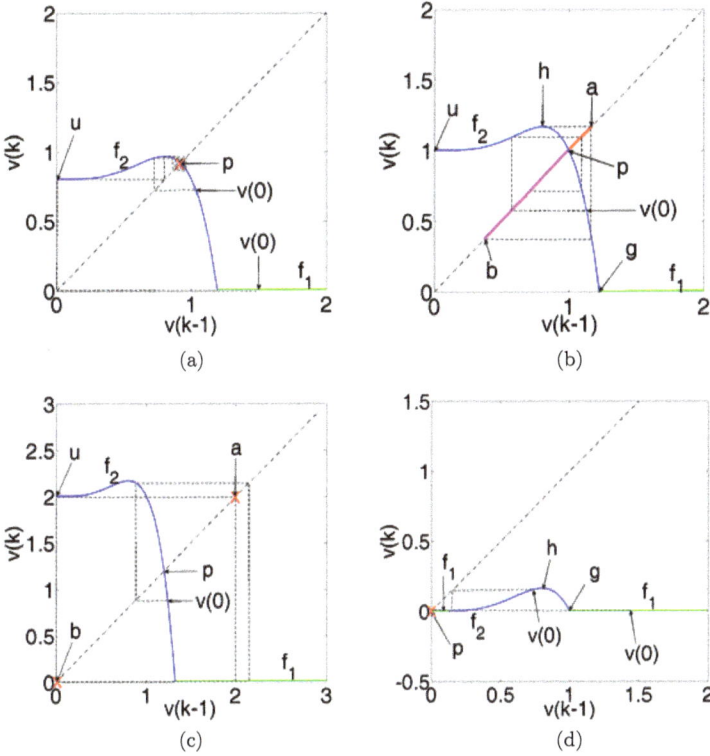

Figure 9.1. Global attractor types of neural firing under pre-critical plasticity: (a) fixed-point, (b) largely-cyclic, (c) oscillatory and (d) silent. As described in the text, cobweb trajectories, initiated at $v(0)$, are represented by black dashed lines and converge to global attractors represented by red and violet line segments or red x. The model is specified by Eqs. (8.11) and (8.12), and the parameter values for each of the global attractors are specified in the text (Baram, 2017a).

multiplexing two sequences from the two intervals, but it may not perform a precise cycle at all (termed then "quasi-cyclic", or "quasi-oscillatory").

(c) *Oscillatory attractor.* For $u = 2$, any initial condition, $v(0)$, will result in convergence of the cobweb trajectory to a cycle alternating between the two points a and b, as depicted in Fig. 9.1(c).

(d) *Silent attractor.* For $u = -0.1$, any initial condition, $v(0)$, will result in convergence of the corresponding cobweb trajectory to the origin, as depicted in Fig. 9.1(d).

It can be seen that the attractor type changes as the activation
level is increased. On the other hand, when the activation becomes
negative, the neuron is silenced. We note that, in contrast to the
developmental stages considered in the following sections, the pre-
critical map Eq. (8.11) does not allow for emergence of chaotic
behavior (specifically, the flatness of f_1 implies that the condition
for chaos cannot materialize under the geometry imposed by the
pre-critical map).

Simulation results

We have simulated the neuronal firing-rate under pre-critical excitatory synaptic plasticity according to Eq. (8.11) for cases (a)–(d)
stated above. Note that for the present model, having $n = 2$, and,
formally, $\tau_m = \tau_\omega = \tau_\theta = 0$, the only parameter left to be determined
is the initial firing-rate, selected as $v_i = 0$. The simulation results are
shown in Fig. 9.2. It can be seen that, as predicted by Fig. 9.1, the
first case (a) yields convergence to a fixed point, case (b) results

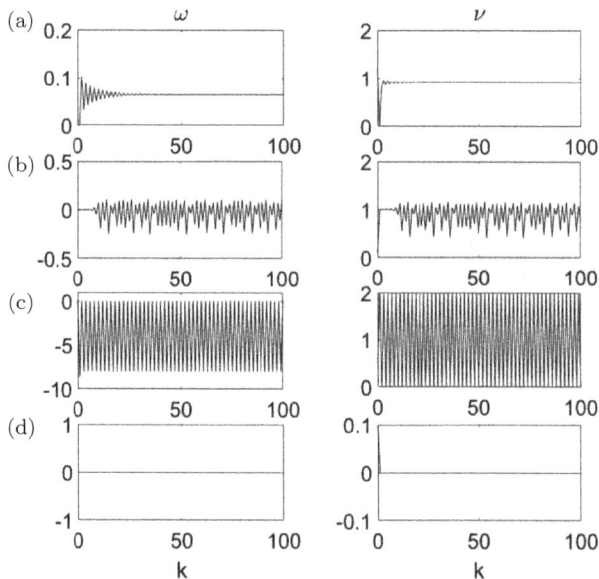

Figure 9.2. Simulated firing modes induced by pre-critical excitatory plasticity
for the models and parameter values specified in the text: (a) fixed point,
(b) quasi-oscillatory, (c) oscillatory, (d) silent (Baram, 2017a).

in quasi-oscillatory behavior, case (c) produces exact oscillation and case (d) yields silence.

9.3 Global attractors of neural firing-rate under critical plasticity

The emergence of the critical plasticity stage, governed by the model Eqs. (8.18)–(8.20), is characterized by an increase in the value of the membrane time constant $\tau_m > 0$, while the synaptic time constants, τ_ω and τ_θ, maintain near-zero values. The nature of the global attractor types is defined by the interplay between λ_1, the slope of f_1, and λ_2, the slope of f_2 at the fixed point p, the intersection point of f_2 with the diagonal $v(k) = v(k-1)$ (note that, for certain parameter values, f_2 may have two additional fixed points neighboring its point of departure from the $v(k)$ axis, the first attracting, the second repelling, as can be envisioned by sliding f_2 downwards in Fig. 9.1(b), with u nearing the origin. Having only a local effect, such occurrence will be omitted in the subsequent analysis). Denoting by x_v and x_h the vertical and the horizontal component of a point x on the map, the global attractor types are specified in the following and depicted in Fig. 9.3 for $n = 2$, $\tau_m = 3$ and $\tau_\omega = \tau_\theta = 0$, and for the representative values of activation (a)$u = 15$, (b) $u = 4$, (c) $u = 1$, (d) $u = -1$.

(a) *Chaotic attractor.* For $u > 0$ and $\lambda_2 < -1$, a chaotic attractor is implied by the condition $q_v \geq p_v$, where p is the fixed point of f_2 and q is the point created by a one-step cobweb sequence initiated at h, as illustrated by Fig. 9.3(a). The attractor is represented in Fig. 9.3(a) by the interval ab, marked by a red line segment on the diagonal $v(k) = v(k-1)$. The period three orbit $a_1 \rightarrow a_2 \rightarrow a_3 \rightarrow a_1$, facilitated by the condition $q_v \geq p_v$, satisfies the Li-Yorke condition for chaos (Li and Yorke, 1975). It was constructed by a numerical search algorithm, which, initiated at the fixed point$p$, takes small steps upwards along the diagonal, each defining a_3 and a corresponding cobweb diagram of length 3. When the diagram closed to a sufficient accuracy, the search was complete.

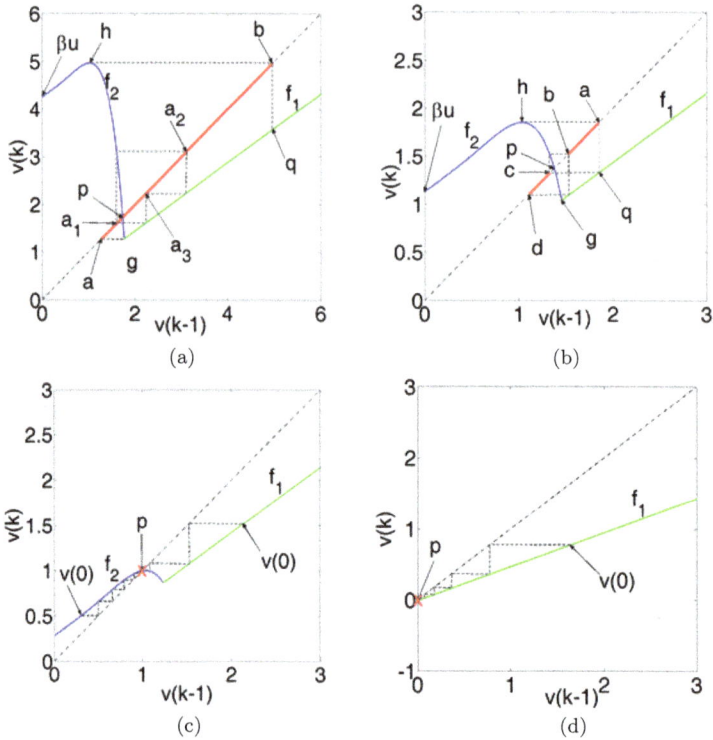

Figure 9.3. Global attractor types under critical plasticity: (a) chaotic, (b) largely-cyclic, (c) fixed-point and (d) silent. Cobweb trajectories, represented by dashed black lines, are defined in the text. Attractors are represented by red line segments, or by points marked by red x, as appropriate, on the diagonal. The model is specified by Eqs. (8.18)–(8.20), and the conditions and parameter values for each of the global attractors are specified in the text (Baram, 2017a).

(b) *Largely-cyclic attractor.* For $u > 0$ and $\lambda_2 < -1$, a largely-cyclic attractor is implied by the condition $q_v < p_v$. The attractor is represented in Fig. 9.3(b) by the intervals ab and cd, separated by the repelling interval bc (note that a, b, c and d are defined by a four-step cobweb sequence initiating at h). A cobweb trajectory may cyclically alternate between the two indicated intervals, possibly performing a precise cycle of even period, multiplexing sequences across the two intervals, but it may not perform a precise cycle at all.

(c) *Fixed-point attractor.* For $u > 0$ and for $\lambda_2 > -1$, we have a fixed point attractor at p. The mode of convergence may be monotone, for $\lambda_2 \geq 0$, as in the case depicted in Fig. 9.3(c), or alternate (increasing $v(k)$ followed by decreasing $v(k+1)$ and vice versa), for $\lambda_2 < 0$.

(d) *Silent attractor.* For $u \leq 0$, the attractor is the origin, as depicted in Fig. 9.3(d). The convergence is monotone, according to f_1, as in Fig. 9.3(d), if $\lambda_1 \geq \lambda_2$ throughout.

Simulation results

We have simulated the neuronal firing-rate under critical plasticity according to Eqs. (8.18)–(8.20), with m satisfying Eq. (8.12), $n = 2$, $\tau_m = 3$, $\tau_w = \tau_\theta = 0$, and zero initial conditions for the following four cases (which were also used in plotting the maps of Fig. 9.3): (a) $u = 15$, (b) $u = 4$, (c) $u = 1$, (d) $u = -1$. The corresponding test parameters for these cases where obtained as:

(a) $\lambda_2 = -13.0830$, $p_v = 1.7339$ and $q_v = 2.6250$, which satisfy the stated conditions for the chaotic attractor type.

(b) $\lambda_2 = -3.4624$, $p_v = 1.3715$ and $q_v = 0.5463$, which satisfy the specified conditions for the largely-cyclic attractor type.

(c) $\lambda_2 = 0.1496$, which satisfies the specified conditions for the fixed-point attractor type.

(d) $\lambda_1 = 0.7165$, the same as λ_2 throughout.

The simulation results for the four cases are depicted in Fig. 9.4, where it can be seen that the firing modes obtained for each of the cases are the ones predicted by the attractor types. Specifically, the persistence of the synaptic weight in cases (a) and (b) is consistent with that of the firing-rate. Case (a) produced a highly disordered sequence, as expected from a chaotic attractor. Case (b) produced a 4-cycle, multiplexing two oscillatory sequence, which might be expected from a largely-cyclic attractor. Cases (c) and (d) resulted in convergence to a constant firing-rate, and to silence, respectively, as predicted by the corresponding global attractor analysis.

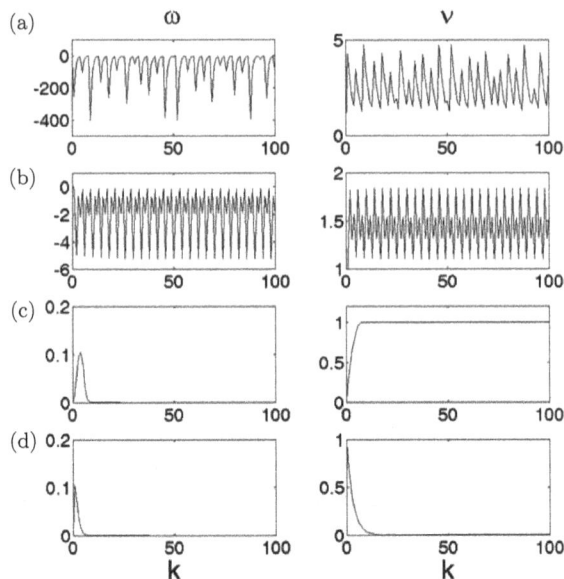

Figure 9.4. Simulated neuronal firing modes under critical plasticity: (a) chaotic, (b) 4-cycle multiplexed, (c) fixed-point, (d) silent. The left column shows time evolution of the feedback synaptic weight, ω, while the right column shows time evolution of the firing-rate, v. Model parameter values are specified in the text (Baram, 2017a).

9.4 Robustness of firing-rate global attractors with respect to model parameter

The variability of the global attractors, or the dynamic modes, of neural firing demonstrated in this book for each of the developmental stages raises the question of the robustness, or, conversely, the sensitivity, of the results with respect to the parameter values. While we have seen that non-regular bifurcations (that is, sharp transitions from one dynamic mode to another, often involving chaotic firing) are caused essentially by changes in the "external" activation u, the time constants more regularly control the nature of the global attractors, or the dynamic modes of firing. Here, we examine, for illustrative purposes, the nature of the changes in the neural firing modes caused by small changes in the single non-zero internal parameter corresponding to the critical stage, namely, the membrane time constant, τ_m. Specifically, we ask whether each of the modes observed is locally robust against small changes in the time constant and whether the modes change in a regular and gradual manner as the time constant changes continuously. For a constant $u = 1$, and for different values of τ_m, gradually changing from 0.1 to 1 in steps of 0.1, then from 1 to 3 in steps of 1, we obtain the cobweb diagrams which, depicted in Figs. 9.5, 9.6 and 9.7, represent the corresponding global attractors. While each of the cobweb trajectories must be initialized at a certain value, $v(0)$, we employ, in certain cases, trajectories

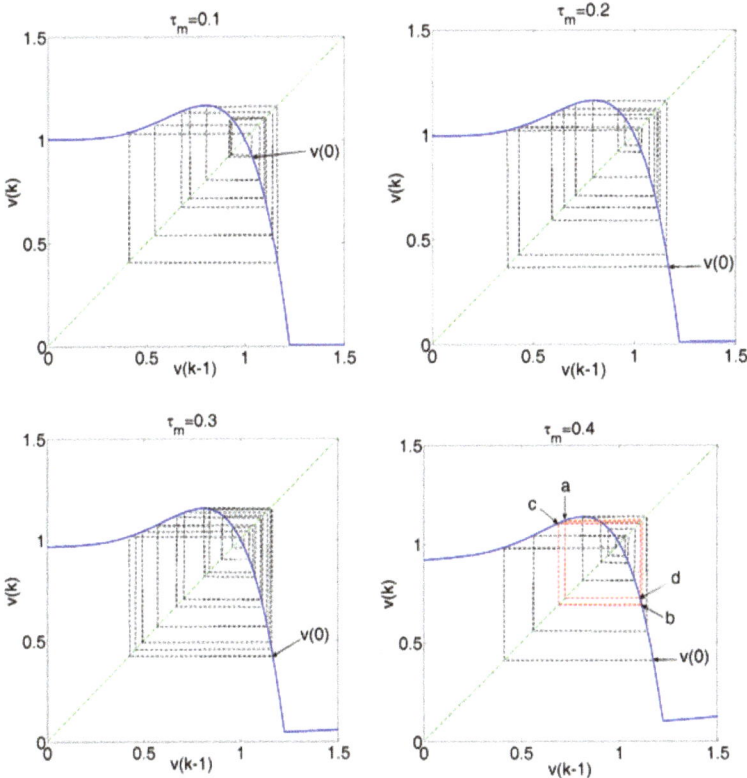

Figure 9.5. As τ_m is varied from 0.1 to 0.2 and then to 0.3, the firing mode, while showing some change, converging to a certain bundle-like pattern, remains largely cyclic. As τ_m attains the value 0.4, the firing mode, as a continuation of its previous convergence, becomes the 4-cycle, represented by the repeated sequence $a - b - c - d$ connected by a red-colored cobweb (Baram, 2017a).

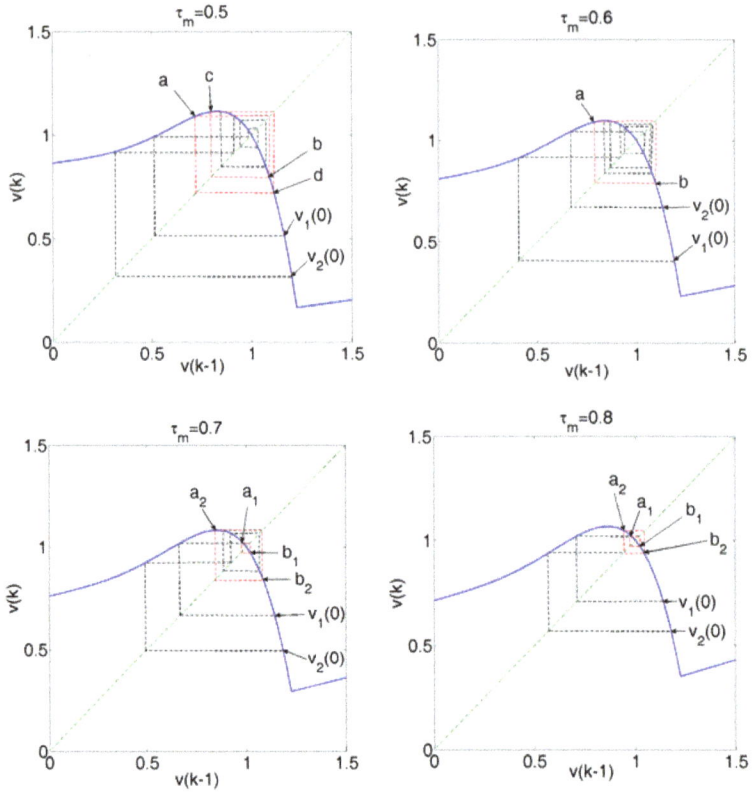

Figure 9.6. As τ_m is varied from 0.5 to 0.8, the global attractor, maintaining its cyclic character, changes from a constant 4-cycle $(a - b - c - d$, for $\tau_m = 0.5)$ to a constant 2-cycle $(a - b$ for $\tau_m = 0.6)$, then becomes an initiation-dependent 2-cycle $(a_1 - b_1$ for $v_1(0)$ and $a_2 - b_2$ for $v_2(0)$, for $\tau_m = 0.7)$, reducing in amplitude for $\tau_m = 0.8$ (Baram, 2017a).

initialized at two different values of $v(0)$ so as to demonstrate the general nature of the firing modes (for instance, for a firing mode which converges to a fixed orbit, all cobweb trajectories, regardless of their initialization points, converge to that orbit). The nature of the resulting firing modes is described as follows:

As τ_m is varied from 0.1 to 0.2 and then to 0.3 (Fig. 9.5), the firing mode, maintaining a largely cyclic character, converges to a bundle-like pattern. As τ_m attains the value 0.4, the firing mode, as a continuation of its previous convergence, becomes a 4-cycle,

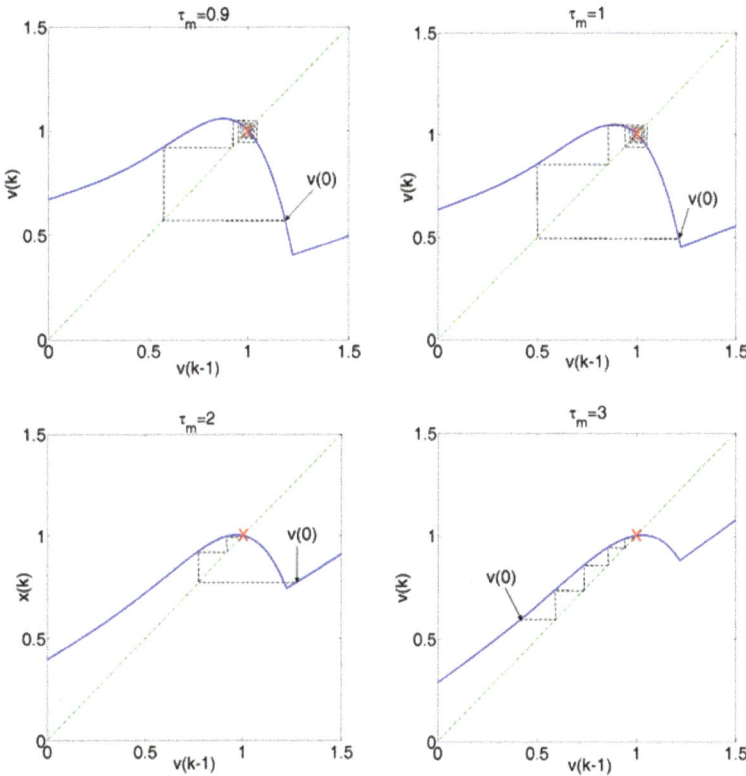

Figure 9.7.   As τ_m continues to increase from 0.8 (Fig. 9.6) to 0.9 (Fig. 9.7), the oscillatory mode continues to reduce in amplitude and converges to a point, represented by a red x. The fixed-point nature of the firing mode remains as τ_m is increased further, while the nature of convergence stiffens (from alternating for $\tau_m = 0.9$ and $\tau_m = 1$ to one-directional for $\tau_m = 2$ and $\tau_m = 3$, Baram, 2017a).

represented by the repeated sequence $a - b - c - d$ connected by a red-colored cobweb.

As τ_m is varied from 0.4 to 0.8 (Fig. 9.6) the firing mode maintains its cyclic character. As τ_m is varied from 0.4 to 0.5, the firing mode remains a 4-cycle, represented by the repeated sequence $a - b - c - d$ connected by a red-colored cobweb. As τ_m is varied from 0.5 to 0.6, the firing mode becomes a 2-cycle, represented by the repeated sequence $a - b$, connected by a red-colored cobweb. For both $\tau_m = 0.5$ and $\tau_m = 0.6$, cobweb trajectories initiated at different initial conditions (here, $v_1(0)$ and $v_2(0)$) converge to the same limit orbit. As τ_m attains the value 0.7, the firing modes corresponding to different initial conditions converge to different 2-cycles, represented by red-colored cobwebs.

For the initial condition $v_1(0)$ the limit oscillatory mode is represented by the repeated sequence $a_1 - b_1$, while for the initial condition v_2 the limit oscillatory mode is represented by the repeated sequence $a_2 - b_2$. As τ_m is varied from 0.7 to 0.8 (Fig. 9.6), the oscillatory modes reduce in amplitude.

As τ_m is varied from 0.8 (Fig. 9.6) to 0.9 (Fig. 9.7), the oscillatory mode continues to reduce in amplitude and converges to a point, represented by a red x at the intersection of the (blue) map and the (green-dashed) line $v(k) = v(k - 1)$. The fixed-point nature of the firing mode remains as τ_m is increased further, (here, to 1, 2 and 3), while the nature of convergence becomes stricter and stricter (from alternating for $\tau_m = 0.9$ and $\tau_m = 1$ to one-directional for $\tau_m = 2$ and $\tau_m = 3$).

We have demonstrated by example, employing a single model parameter, that global attractors (or dynamic modes) of neural firing-rate are robust with respect to the parameter. Small changes in the parameter value cause small changes in the firing mode, while any change in the nature of the firing mode is moderate and gradual.

9.5   Discussion

Employing cobweb diagrams, we have characterized the global attractors of synaptically-modulated firing-rates corresponding to early development with visual clarity, ratified by sequential simulation. The differences between the discrete iteration maps corresponding to the pre-critical and the critical stages are vividly demonstrated by the corresponding cobweb diagrams in Figs. 9.1 and 9.3, respectively. While, as can be seen from the two figures, different parameters produce different global attractors within the corresponding domains, there are certain parameter-independent characteristic differences between the global attractor repertoires of the two stages. In particular, while the inherent nullification of the f_1 part of the map corresponding to the pre-critical stage (Fig. 9.1) does not allow for a chaotic mode of firing-rate to evolve, the positive slope of the f_1 part of the map corresponding to the critical stage (Fig. 9.2) does facilitate chaotic behavior (Figs. 9.3(a) and 9.4(a)). While a flat initiation of the f_2 part of the map corresponding to the pre-critical stage (Fig. 9.1(a)) represents a decayed excitatory behavior of the synaptic weight ω (Fig. 9.2 (a)), an increased positive slope initiation of the f_2 part of the map (Figs. 9.1 (b) and (c)) produces persistent excitation of the synaptic weight ω (Figs. 9.2 (b) and (c)). Similarly, the positive slope of the f_2 part of the map corresponding to the critical stage, mirrored by the positive slope of the f_1 part of the map (Figs. 9.3(a) and (b)) produces persistent excitation of the synaptic weight ω, as evidenced in cases (a) and (b) of Fig. 9.4. Indeed, the early developmental stages have been associated in empirical studies with "persistent plasticity" (e.g., Constantine-Paton *et al.*, 1990; Hensch, 2005; Miyata *et al.*, 2012).

Chapter 10

Global Attractors of Neural Firing-rate in Maturity

10.1 Introduction

As we have noted earlier, synaptic maturity, characterized by non-zero membrane and synapse time constants, τ_m, τ_ω and τ_θ, is manifested by convergent synaptic weights as represented by Eqs. (8.22)–(8.24). Here, we show that the *bilinear threshold* structure of the map facilitates a formal global attractor specification and categorization in considerable analytic detail. The global attractor of the map is found to morph into 12 attractor types, which we call the *global attractor alphabet* of neural firing modes. For analytic convenience, the global attractor types are categorized into two groups, the first characterized by negative (inhibitory) λ_2, the second by positive (excitatory) λ_2. While the first group corresponds to positive activation, the second is subdivided into (i) an active attractor group driven by positive activation and (ii) a passive attractor group driven by negative activation. A change in the sign of activation will transform an active attractor into a passive one (concealment), or a passive attractor into an active one (revelation). In the case of spontaneous time-dependent activation variation (Connor and Stevens, 1971), such transformation may in itself become a secondary dynamic mode. A case in point is periodic bursting (Elson *et al.*, 2002) instigated by post-inhibitory rebound (Perkel and Mulloney, 1974).

10.2   Global attractors of neural firing-rate
in maturity

A key determinant of the dynamic nature of a global attractor in maturity is the slope λ_2 of the function f_2, as represented by Eq. (8.22). As can be seen from Eq. (8.24), λ_2 can be positive or negative, depending on the sign of the limit synaptic weight, ω, and the circuit size, n. For analytic and organizational convenience we divide the global attractors of maturity into two groups, the first associated with $\lambda_2 < 0$, the second with $\lambda_2 \geq 0$. In the first case we will also assume that the activation is positive ($u > 0$), otherwise, as can be readily verified, the map has a single attractor at the origin. In the second case ($\lambda_2 \geq 0$), we will consider both positive and negative values of u.

Group 1: $\lambda_2 < 0, u > 0$

(a) *Chaotic attractor:* The Li-Yorke theorem (Li and Yorke, 1975) states that a map f is chaotic if there exists x such that $x \neq f(x)$ and $x = f^3(x) = f(f(f(x)))$. The latter equation is said to describe an orbit of period 3. In Fig. 10.1(a), an orbit of period 3 is represented by $a_1 \to a_2 \to a_3 \to a_1$. In the space spanned by the coordinates $v(k-1)$ and $v(k)$, the line $v(k) = f_1(v(k-1))$, having a positive slope smaller than 1, meets the diagonal $v(k) = v(k-1)$ only at the origin. On the other hand, when the line $v(k) = f_2(v(k-1))$ intersects the diagonal, it will be at

$$p = \frac{\beta u}{1 - \beta n \omega - \alpha} \tag{10.1}$$

The latter is, then, the only possible fixed point of the scalar map at hand, beside the origin. The point q in Fig. 10.1(a) is defined by a three-step cobweb trajectory initiated at the bend-point g.

Suppose first that $u > 0$ and

$$\lambda_2 < -1/\lambda_1 \tag{10.2}$$

(hence, $\lambda_2 < -1$, as $0 > \lambda_1 < 1$). Let p_v and q_v denote the vertical coordinates of points p and q, respectively, on the map in Fig. 10.1(a))

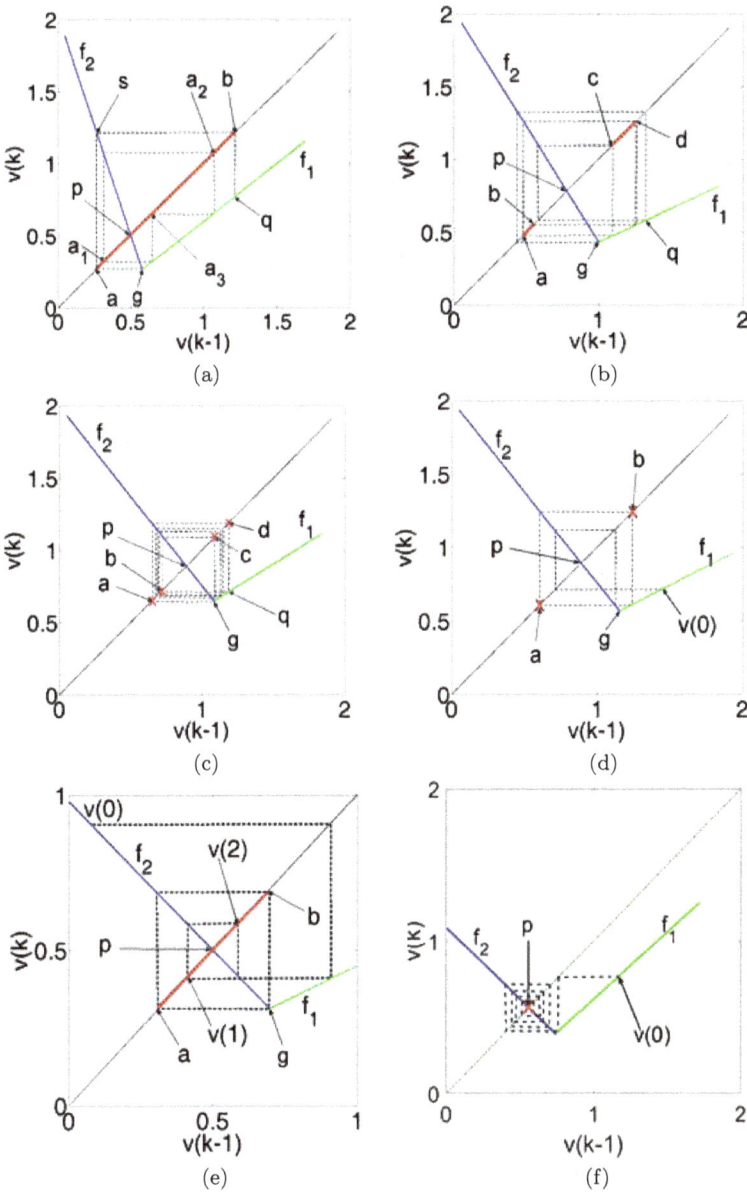

Figure 10.1. Group 1 global attractors of firing-rate in maturity: (a) chaotic attractor (b) largely oscillatory attractor (c) multiplexed oscillation attractor, (d) fixed oscillation attractor, (e) initial condition-dependent oscillatory attractor, (f) alternate fixed-point attractor.

and consider the domain:

$$q_v \geq p_v \tag{10.3}$$

It can be readily verified that if condition 10.2 is satisfied, then condition 10.3 is necessary and sufficient for the existence of an orbit of period 3, hence, for the existence of chaos. It follows that a sufficient condition for chaos is that both conditions 10.2 and 10.3 are satisfied.

Some calculations will show that

$$q_v = \left(\frac{\lambda_1 \lambda_2}{\lambda_1 - \lambda_2} + 1\right) \lambda_1 u / \tau_m \tag{10.4}$$

and

$$p_v = \frac{u}{\tau_m(1 - \lambda_2)} \tag{10.5}$$

which implies that condition 10.3 is equivalent to the condition

$$\lambda_2 \leq \frac{-1 - \sqrt{1 + 4\lambda_1^2}}{2\lambda_1} \tag{10.6}$$

Denoting

$$c_1 = 2\lambda_1 \lambda_2 + 1 + \sqrt{1 + 4\lambda_1^2} \tag{10.7}$$

and

$$c_2 = \lambda_1 \lambda_2 + 1 \tag{10.8}$$

It can be seen that condition 10.6 (hence, condition 10.3) can be written as the condition $c_1 \leq 0$, while condition 10.2 can be written as the condition $c_2 < 1$. For $u > 0$, a sufficient condition for chaos is then $c_1 \leq 0$ and $c_2 < 0$, yielding condition 10.3.

(b) *Largely oscillatory attractor:* It follows that if, while condition 10.2 is satisfied, the following condition

$$q_v < p_v \tag{10.9}$$

is satisfied, the map under consideration (Eqs. (8.22)–(8.24)) cannot be chaotic.

Condition 10.9 is equivalent to the condition

$$\frac{-1 - \sqrt{1 + 4\lambda_1^2}}{2\lambda_1} < \lambda_2 < \frac{-1 + \sqrt{1 + 4\lambda_1^2}}{2\lambda_1} \tag{10.10}$$

which, combined with condition 10.2, is equivalent to the condition

$$\frac{-1 - \sqrt{1 + 4\lambda_1^2}}{2\lambda_1} < \lambda_2 < -1/\lambda_1 \tag{10.11}$$

A global attractor which satisfies condition 10.11 consists of two alternating intervals $[a, b]$ and $[c, d]$, separated by the repelling interval $[b, c]$, defining case (b), illustrated in Fig. 10.1(b). As can be geometrically verified, there cannot be an orbit of period 3 in this case. Hence, in contrast to the global attractor presented in Fig. 10.1(a), this global attractor in Fig. 10.1(b) is not chaotic. It can be characterized as a multiplexity of the two mutually exclusive intervals $[a, b]$ and $[c, d]$. Resembling the multiplexing of temporal sequences, often used for efficient communication in technological systems, the *interval multiplexity* at hand implies a certain degree of ordering, imposed by alternate motions along the two branches f_1 and f_2 of the map. Since these motions do not constitute exact oscillations, they may be characterized as "largely oscillatory".

(c) *Multiplexed oscillation attractor:* It can be verified that for $-1/\lambda_1 \leq \lambda_2 \leq -1$, the sequence generated by the map Eq. (8.22), initiated, say, on the right-side of the bend-point g in Fig. 10.1 (c), consists of the following two subsequences:

$$v(2k) = f_1 \circ f_2(v(2(k-1)) = \lambda_1\lambda_2 v(2(k-1)) + \lambda_1 u/\tau_m; \quad v(0) = v_0 \tag{10.12}$$

and

$$v(2k+1) = f_2 \circ f_1(v(2(k-1)) = \lambda_1\lambda_2 v(2k-1) + u/\tau_m;$$
$$v(1) = f_2(v_0) = \lambda_2 v_0 + u/\tau_m \tag{10.13}$$

For $\lambda_2 = -1/\lambda_1$, yielding $\lambda_1\lambda_2 = -1$, Eqs. (10.12) and (10.13), respectively, yield two multiplexed oscillations

$$v(2k) = -v(2(k-1)) + \lambda_1 u/\tau_m; \quad v(0) = v_0 \tag{10.14}$$

and

$$v(2k+1) = -v(2k-1) + u/\tau_m; \quad v(1) = f_2(v_0) = \lambda_2 v_0 + u/\tau_m$$
$$(10.15)$$

as depicted in Fig. 10.1(c).

(d) *Fixed oscillation attractor:* For $-1/\lambda_1 < \lambda_2 < -1$, yielding $-1 < \lambda_1\lambda_2 < -1$, the subsequence Eq. (10.12) converges to

$$a = u/\tau_m(1 - \lambda_1\lambda_2) \tag{10.16}$$

while the subsequence 10.13 converges to

$$b = \lambda_1 u/\tau_m(1 - \lambda_1\lambda_2) \tag{10.17}$$

The firing-rate trajectory converges, then, to an oscillation between the two values, a and b, as illustrated by Fig. 10.1(d).

(e) *Initial condition-dependent oscillatory attractor:* The case $\lambda_2 = -1$, depicted by Fig. 10.1 (e), is characterized by the bend point g of the map. Any cobweb trajectory will be attracted to the dash-dotted square shown and any cobweb trajectory entering that domain will form a perfect square within it. Equivalently, any trajectory entering the interval $[a, b]$ at a point v will form an oscillation between v and $2q_p - v$. The interval $[a, b]$ is, then, a global attractor, each point of which is of period 2.

(f) *Alternate fixed-point attractor:* For the case $-1 < \lambda_2 < 0$, depicted by Fig. 10.1(f), we have a fixed point attractor at p, enforcing alternate convergence (an increase in the value of $v(k)$ followed by a decrease, and vice versa).

Group 2: $\lambda_2 \geq 0$

    Figure 10.2 displays global attractors of firing-rate for $\lambda_2 \geq 0$. All cases show point attractors, which are categorized as non-zero fixed-point attractors, attractors at zero, attractors at infinity, or a bi-polar attractor at zero and infinity. The mode of convergence depends on the correspondence between λ_1 and λ_2, the slopes of f_1 and f_2, respectively. The attractors of this group are specified below:

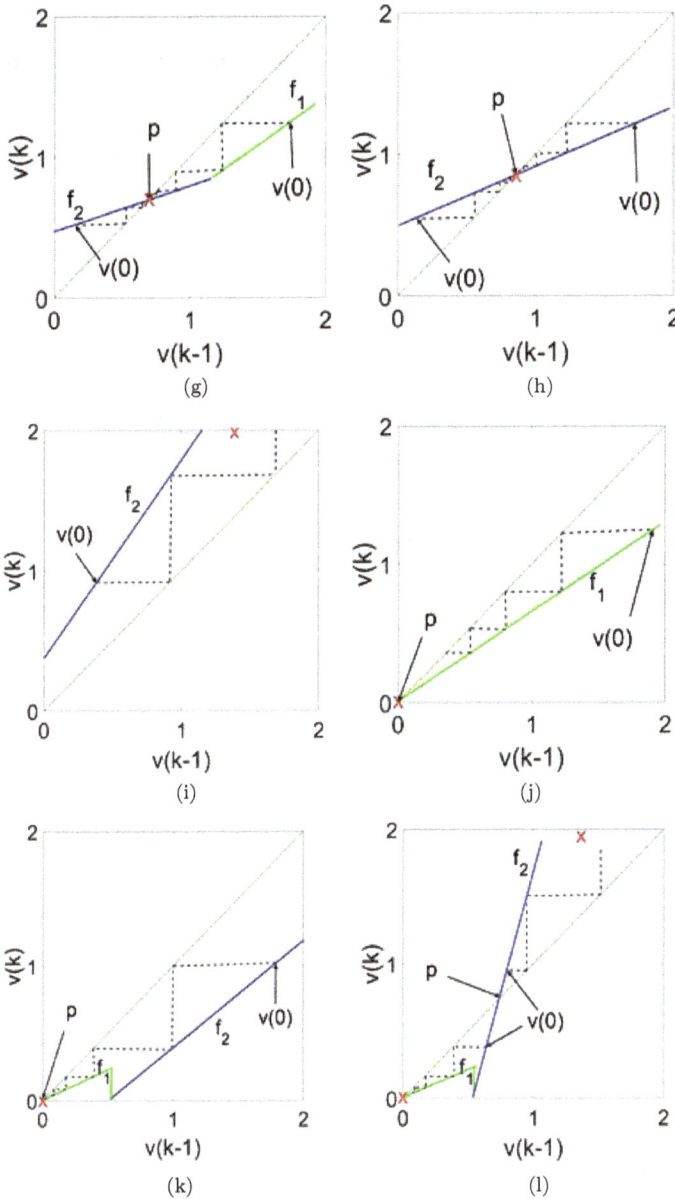

Figure 10.2. Group 2 global attractors of firing-rate in maturity: (g) bi-modal fixed point, (h) unimodal fixed-point (i) attractor at infinity, (j) unimodal attractor at zero, (k) bi-modal attractor at zero, (l) bi-polar attractor at zero and infinity. (l) For $u < 0$ and $\lambda_2 > 1$, Fig. 10.2(l) shows a *bipolar attractor at zero and infinity*, with p a repelling fixed-point.

(g) For $u > 0$ and $0 < \lambda_2 < \lambda_1$, Fig. 10.2(g) shows a *fixed-point attractor* at p inducing *bi-modal convergence* corresponding to a transition from f_1 to f_2.

(h) For $u > 0$ and $\lambda_2 > \lambda_1$, Fig. 10.2(h) shows a *fixed-point attractor* at p inducing a *unimodal convergence* mode corresponding to f_2.

(i) For $u > 0$ and $\lambda_2 > 1$, Fig. 10.2(i) shows an *attractor at infinity*.

(j) For $u \leq 0$ and $\lambda_2 \leq \lambda_1$, Fig. 10.2(j) shows an *attractor at zero* inducing *unimodal convergence* mode corresponding to f_1.

(k) For $u < 0$ and $\lambda_2 > \lambda_1$, Fig. 10.2(k) shows an *attractor at zero* inducing *bimodal convergence* mode corresponding to transition from f_2 to f_1.

(l) For $u < 0$ and $\lambda_2 > 1$ the attractor is bipolar at zero and infinity. The final destination of a trajectory at zero (silence) or infinity (saturation) is determined by the initial condition with p being the point of separation between the two basins.

10.3   Simulation results

We have simulated the model Eqs. (8.4), (8.6) and (8.7) for the attractor types (a)–(e) earlier, for $n = 2$ and the parameter values specified in the following. For each of the cases, the steady-state value of the synaptic weight, w, was calculated by driving Eqs. (8.4), (8.6) and (8.7), with $w(0) = 0$ and $v(0) = 1$, to convergence (practically, this was achieved for $N = 100$), and the corresponding values of $\lambda_1, \lambda_2, c_1$ and c_2 were calculated by Eqs. (8.23), (8.24), (10.7) and (10.8), respectively.

(a) $u = 20, \tau_m = 2.5, \tau_w = 1000, \tau_\theta = 2$, yielding

$$w = -6.7711, \lambda_1 = 0.6703, \lambda_2 = -3.7943,$$

$$c_1 = -2.4143, \ c_2 = -1.5434$$

(b) $u = 20$, $\tau_m = 4$, $\tau_\omega = 200$, $\tau_\theta = 20$, yielding

$$\omega = -5.1502, \lambda_1 = 0.7788, \lambda_2 = -1.4996,$$
$$c_1 = 0.5151, \ c_2 = -0.1679$$

(c) $u = 4$, $\tau_m = 2$, $\tau_\omega = 300$, $\tau_\theta = 0.1$, yielding

$$\omega = -2.5695, \lambda_1 = 0.6065, \lambda_2 = -1.4155,$$
$$c_1 = 0.8550, \ c_2 = 0.1415$$

(d) $u = 1$, $\tau_m = 2$, $\tau_\omega = 5$, $\tau_\theta = 1$, yielding

$$\omega = -0.1925, \lambda_1 = 0.6065, \lambda_2 = 0.5308,$$
$$c_1 = 3.2160, \ c_2 = 1.3219$$

(e) $u = -1$, $\tau_m = 2$, $\tau_\omega = 5$, $\tau_\theta = 1$, yielding

$$\omega = 0, \lambda_1 = 0.6065, \lambda_2 = 0.6065, \quad c_1 = 3.3079, c_2 = 1.3679$$

It can be seen from the parameter values and the condition numbers specified above that cases (a), (b) and (c) of the simulation correspond to cases (a), (b) and (c) of group 1, associated with chaotic, largely oscillatory and oscillatory attractors represented by Figs. 10.1(a), 10.1(b) and 10.1(c), while cases (d) and (e) of the simulation correspond to cases (g) and (j) of Group 2, associated with bimodal fixed-point attractor and attractor at zero, represented by Figs. 10.2(g) and 10.2(j) respectively.

Each test consisted of a learning stage, in which the feedback synaptic weight evolved from initial to final value, and a retrieval stage, in which the final value of the feedback synaptic weight obtained in the learning stage was used. The simulation results are shown in Fig. 10.3, where the values of the feedback synaptic weight throughout the run, the firing-rate during learning throughout the run, the firing-rate during learning in the late period of the run, and

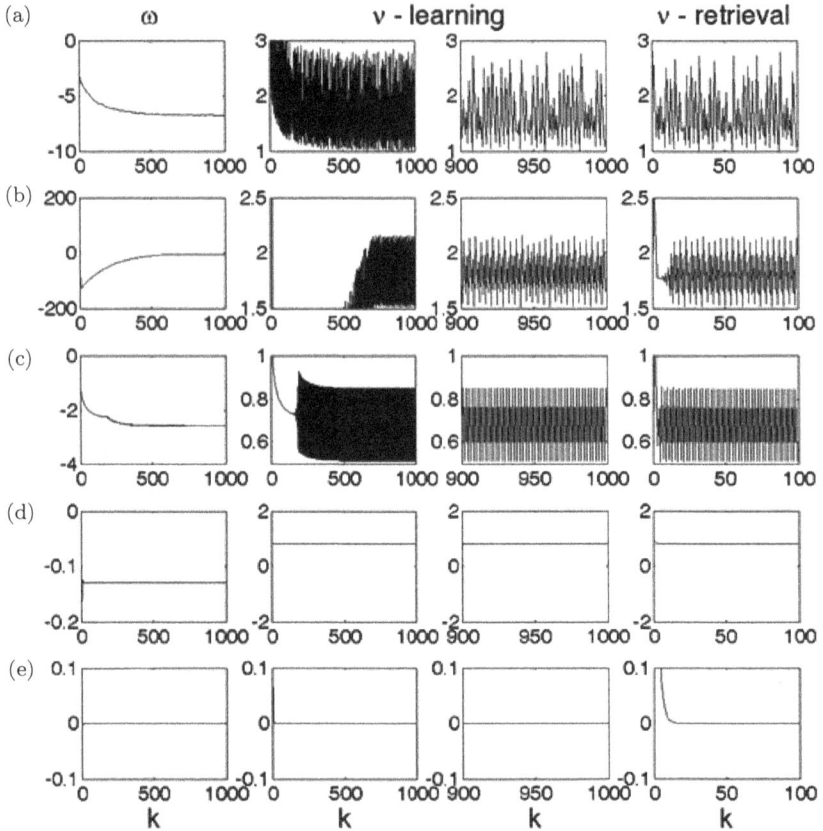

Figure 10.3. Simulated sequences under synaptic maturity during learning and retrieval, representing: (a) chaotic, (b) largely-cyclic, (c) oscillatory, (d) fixed-point and (e) silent modes. The first column represents synaptic weight (ω) evolution throughout the run, while the remaining three represent the firing-rate (v) during learning throughout the run and in the late period of the run, and during retrieval in the early period of the run, respectively (Baram, 2017a).

the firing-rate during retrieval in the early period of the run are given, respectively, in the four columns. It can be seen that, in each of the cases, the feedback synaptic weight converged to a constant value. It can also be seen that, while convergence in retrieval is noticeably faster than convergence in learning for the first three cases, learning and retrieval end, as desirable, at the same attractor. The convergent modes of firing visually conform to the attractor types predicted by the theory (and illustrated by Figs. 10.1 and 10.2). Specifically, case (a) produced a highly disordered, bifurcated and bursting sequence, as expected from a chaotic attractor, case (b) produced a quasi-4-cycle, multiplexing two uneven 2-cycles, and case (c) produced an oscillatory sequence, while cases (d) and (g) produced fixed and silent sequences, respectively.

10.4   Discussion

The bilinear threshold structure of the rectified neural firing-rate model and the convergent synaptic weights facilitate effective analytic characterization of the global attractors of neural firing-rate in maturity. The conditions for each of the global attractors are specified in terms of the parameters of the corresponding discrete iteration map. This does not only provide an analytic specification of the global attractor, but also a highly intuitive graphical description of the attractor, hence, the corresponding dynamic mode. Specifically, the parameters λ_1 and λ_2, which define the slopes of a map, also define the bifurcation points of the global attractors. These parameters, having a direct geometric manifestation, can be highly valuable in the mathematical analysis of the global attractors. For instance, in view of the often reported inadequacy of empirical indicators of chaos (Sprott, 2003), such as the first Lyapunov exponent (Wright, 1984) used in the analysis of empirical neurophysiological data (Fell *et al.*, 1993), the indicators c_1 (Eq. 10.7) and c_2 (Eq. 10.8), directly related to λ_1 and λ_2, analytically and unambiguously define the boundaries of chaotic attractor domain. Similarly, the parameters λ_1 and λ_2 analytically define the domains of largely-oscillatory attractor, the multiplexed oscillation attractor and the fixed oscillation attractor. Other attractors, more simply structured, were identified by the signs of λ_2 and the activation u. The ability to characterize the modes of firing-rate dynamics in such analytic detail appears to have been only recently noted (Baram, 2013a).

Similarly behaved biological firing modes, notably, chaotic (Hayashi and Ishizuka, 1992; Fell *et al.*, 1993), bursting (Hayashi and Ishizuka, 1992), quasi-cyclic (Lankheet *et al.*, 2012), oscillatory (Ditlevsen and Locherbach, 2017), multiplexed (Panzeri *et al.*, 2009), fixed point, or tonic (Bennet *et al.*, 2000), and silent (Melnick, 1994; Atwood and Wojtowicz, 1999) have been reported. While chaos has been identified with ambiguity and deteriorated performance in certain applications involving artificial feedforward networks (Bertels *et al.*, 1998; Gicquel *et al.*, 1998; Reco-Martinez *et al.*, 20005; Verschure, 1991) and in biological neural networks (Fell *et al.*, 1993), several works have noted the utility of chaos in learning and problem solving by annealing (Lin, 2001; Ohta, 2002; Liu *et al.*, 2006; Verschure, 1991; Sato *et al.*, 2002). Moreover, it has been claimed (Langton, 1990) and disputed (Mitchell *et al.*, 1993), that the boundary between order and chaos provides favorable conditions for universal computation. Chaos has been incorporated into neuron models through a bimodal logistic function (Aihara *et al.*, 1990), simplified phenomenological models (Lin *et al.*, 2002; Shilnikov and Rulkov, 2003), or negative self-feedback (Lin, 2001; Ohta, 2002; Liu *et al.*, 2006). However, the problem solving and computation capabilities of biological neural networks are rather vague notions, and the possible role of chaos in such networks has remained a mystery. The analytic approach taken in this work is particularly noteworthy in view of the fact that empirical measures, such as the Lyapunov exponents, often fail to provide conclusive evidence of chaos. We suggest that the highly patterned geometric and statistical structures of the chaotic attractors associated with neural networks make such attractors an integral part of the neural code. While our mathematical analysis has primarily led to the consideration of symmetric neural circuits, resulting, as in the cases considered by Field and Golubitsky (2009), in symmetric chaotic attractors, we have also noticed in simulation that highly patterned chaotic attractors arise in non-symmetric neural circuits. Providing a transcendental prescription for mixing different firing-rates, a chaotic attractor constitutes a natural mechanism for statistical encoding and temporal multiplexing of neural information. Signal multiplexing is known to enhance information transmission capacity and is widely used in communication systems (Li and Stuber, 2010). However, in contrast to technological applications, which require rhythmic and synchronous multiplexing, especially for the purpose of decoding (Keller *et al.*, 2001), biological multiplexing does not appear to require temporal precision (Panzeri *et al.*, 2009). It should be noted that the application of chaos for the purpose of synchronization in technological communication systems (Itoh and Chua, 1997) is quite different from the present context, where synchronization is not a pre-requisite. Since the model considered in the present work deals with firing-rates, amplitude corresponds to firing frequency. Depending on the function of the receiving neuron, demultiplexing can be done, in principle, by bend-pass filtering. Neuronal low-pass (Pattersen and Einevoll, 2008) and high-pass (Poon *et al.*, 2000) filtering have been reported. Yet, the raw multiplexed signal, representing, as implied by the models under consideration, locally smoothed (averaged) information, can be useful in sensory systems (multiplexed Red, Green and Blue (RGB) color coding is a known example found in both biological and technological vision systems (Hunt, 2004)).

Chapter 11

Firing-rate Mode Segregation by Neural Circuit Polarization

11.1 Introduction

The Hebb paradigm (Hebb, 1949; Hertz *et al.*, 1991), supported by the eigen-frequency preference paradigm of spiking neurons (Izhikevich, 2001), implies inhibitory effects between asynchronously firing neurons. This will regularly tend to segregate a circuit into internally-synchronous, externally-asynchronous subcircuits. However, as we show in this chapter, the resulting mutually asynchronous firing of the segregated subcircuits is likely to result in mutual interference between the corresponding firing-rate modes. On the other hand, as we show, the segregation of internally-synchronous externally-asynchronous subcircuits by synapse silencing results in interference-free firing. It should be noted that the analysis of asynchronous circuit firing-rate dynamics cannot employ cobweb diagrams, which are essentially limited to scalar systems. We therefore employ representative simulation of small circuits. While the simulation of circuits of many asynchronous neurons would require prohibitive numbers of simulation plots, small circuits are rather effective in demonstrating the main concepts of interest.

As in the case of asynchronous circuits, the segregation of synchronous circuits into internally synchronous, externally asynchronous subcircuits is achieved by synaptic silencing. In contrast to asynchronous circuits, the dynamics of synchronous circuits can be described by cobweb diagrams. The segregation of synchronous circuits, consisting of identical neurons firing in concert, has a somewhat surprising, yet explainable effect: although the neurons remain the same, their firing-rate dynamics change. This can facilitate the simultaneous execution of different cortical tasks.

11.2   Dynamics of somatic and synaptic silencing

The polarity of the individual neuron's somatic membrane potential with respect to the membrane activation threshold is identical to the sign of the operand of the function f in Eq. (7.4). A negative operand will imply that the map Eq. (7.4) takes the form

$$v(k) = \alpha v(k - 1) \tag{11.1}$$

which, as $0 < \alpha < 1$, converges to 0 as $k \to \infty$. It further follows from Eq. (7.6) that, as $v(k)$ converges to 0, so does $w(k)$. Membrane silencing in the individual neuron implies, then, nullification of the feedback synaptic weight.

Synaptic polarity in the individual neuron is controlled by the value of the corresponding synaptic weight. Synapse silencing, represented by a zero value of the feedback synaptic weight, accommodates a positive constant value of u, which, by Eq. (7.4), yields the map

$$v(k) = \alpha v(k - 1) + \beta \, \mathrm{u} \tag{11.2}$$

Note that, since $\alpha < 1$, the line represented by Eq. (11.2) can only intersect the diagonal

$$v(k) = v(k - 1) \text{ if, indeed}, u > 0.$$

As noted before, scalar global attractors are graphically described by cobweb diagrams. The cobweb diagram depicted in Fig. 11.1(a), corresponding to the parameter values $\tau_m = 2$, $\tau_w = 300$, $\tau_\theta = 0.1$ and $u = -1$, illustrates the membrane silencing process, converging to the point p represented by a red X placed at the origin (note that, by Eqs. (8.6) and 11.1, w becomes 0, regardless of the values of τ_w and τ_θ). The cobweb diagram depicted in Fig. 11.1(b), corresponding to the parameters $\tau_m = 2$ and $u = 1$, illustrates the synapse silencing

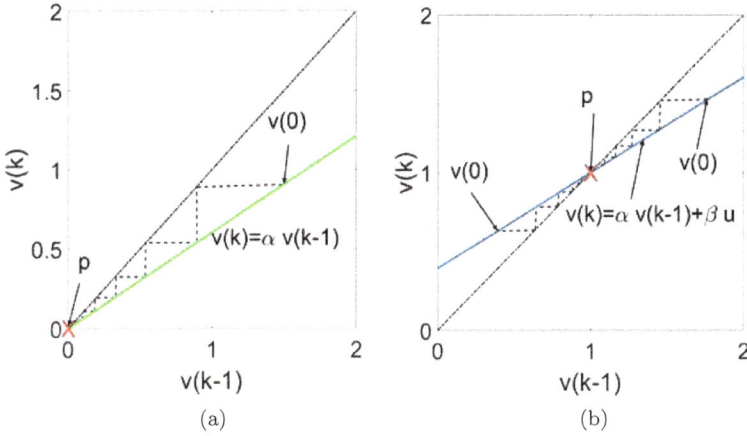

Figure 11.1. Membrane and synapse silencing in the individual neuron. While membrane silencing is represented by a global attractor at the origin (a), feedback synapse silencing is represented by a non-zero fixed-point attractor (b). While membrane silencing stops neuron firing altogether, synapse silencing allows for continued firing due to external activation (Baram, 2018).

process, converging to the global attractor at point p. It follows then that in contrast to membrane silencing, synapse silencing, even in the case of the single feedback synapse of the individual neuron, allows for continued neuronal firing due to external activation.

Membrane or synapse silencing is one side of local polarization. The other side is membrane or synapse activation. These states, governed by Eqs. (7.4) and (7.6), respectively, will affect the modes of the firing-rate dynamics, as discussed next.

11.3   Underlying asynchronous firing-rate and plasticity models

The underlying instantaneous asynchronous firing-rate model is obtained from Eq. (8.4) as

$$v_i(k) = \alpha_i v_i(k-1) + \beta_i f(\boldsymbol{w}_i^T(k)\boldsymbol{v}(k-1) + u_i) \qquad (11.3)$$

where $\boldsymbol{v}(k)$ is the vector of the circuit neurons' firing-rates, v_i, $i = 1, 2, \ldots, n$, $\alpha_i = \exp(-1/\tau_{m_i})$ and $\beta_i = 1-\alpha_i$, with τ_{m_i} the membrane time constant of the i'th neuron, $\boldsymbol{w}_i(k)$ is the vector of synaptic weights corresponding to the circuit's pre-neurons of the i'th neuron (including self-feedback), $u_i = I_i - r_i$, with I_i the circuit-external activation input and r_i the membrane resting potential of the i'th neuron, and f is the conductance-based rectification kernel defined by Eq. (2.16).

The Bienenstock-Cooper-Munro plasticity rule (Bienenstock *et al.*, 1982), Eq. (7.6) now takes the multi-neuron discrete-time form

$$\boldsymbol{w}_i(k) = \varepsilon_i \boldsymbol{w}_i(k-1) + \gamma_i [v_i(k-1) - \theta_i(k-1)] \, \boldsymbol{v}^2(k-1) \quad (11.4)$$

where $\boldsymbol{v}^2$ is the vector whose components are the squares of the components of $\boldsymbol{v}$, $\varepsilon_i = \exp(-1/\tau_{w_i})$, $\gamma_i = 1 - \varepsilon_i$ and Eq. (7.7) now takes the form

$$\theta_i(k) = \delta_i \sum_{\ell=0}^{N} \exp(-\ell/\tau_{\theta_i}) v_i^2(k-\ell) \quad (11.5)$$

where $\delta_i = 1/\tau_{\theta_i}$. In the following sections, we analyze and demonstrate the effects of membrane and synapse polarization on circuit structure and firing dynamics.

11.4 Asynchronous synaptic polarity-gated firing mode segregation

In order to demonstrate the difference between Hebbian segregation and polarity-gated segregation of asynchronous neural circuits, we consider a three-neuron circuit where the first two neurons have the same parameter values and the third has different parameter values, as specified below:

$$u_1 = u_2 = 1, u_3 = 5, \ \tau_{m,1} = \tau_{m,2} = 1, \ \tau_{m,3} = 3, \ \tau_{w,1} = \tau_{w,2} = 5,$$

$$\tau_{w,3} = 0.5, \ \tau_{\theta,1} = \tau_{\theta,2} = 1, \ \tau_{\theta,3} = 0.5$$

Starting with full connectivity, changes in circuit connectivity due to synapse silencing are illustrated in Fig. 11.2. The resulting changes in the neuronal firing modes, simulated by running Eqs. (11.3)–(11.5), are displayed in Fig. 11.3.

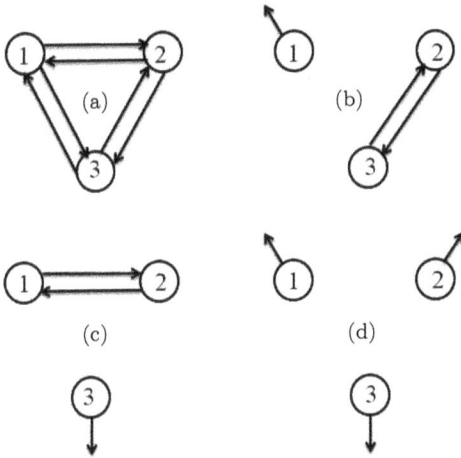

Figure 11.2. Asynchronous 3-neuron circuit segregation by synapse silencing. (a) Fully connected circuit. (b) Asynchronous circuit segregation into a single isolated neuron and an asynchronous 2-neuron circuit by synapse silencing. (c) Asynchronous circuit segregation into a synchronous 2-neuron circuit and a single neuron by synapse silencing. (d) Asynchronous circuit segregation into three single neurons by synapse silencing (Baram, 2018).

The Hebb paradigm, if fully applicable, should segregate the fully connected three-neuron circuit into two totally isolated subcircuits, the first consisting of the two synchronous neurons (neurons 1 and 2) firing in unison according to one firing-rate mode, and the second consisting of one neuron (neuron 3), firing in a different firing-rate mode. Yet, as can be seen in Fig. 11.3(a), there is visible mutual interference between the firing modes corresponding to the fully connected circuit of Fig. 11.2(a). Isolating neuron 1 by polarity-gated synapse silencing (as in Fig. 11.2(b)), it produces a smooth fixed-point firing-rate mode, while neurons 2 and 3 continue to interfere with each other, as can be seen in Fig. 11.3(b). When neuron 3 is isolated from neurons 1 and 2 by polarity-gated synapse silencing (as in Fig. 11.2(c)), the latter form a 2-neuron circuit, which fires in fully matched fixed-point modes, while neuron 3 fires in an oscillatory mode, as can be seen in Fig. 11.3(c). When each of the neurons is isolated by polarity-gated synapse silencing (as in Fig. 11.2(d)), each fires in its characteristic mode, as can be seen in Fig. 11.3(d). It can

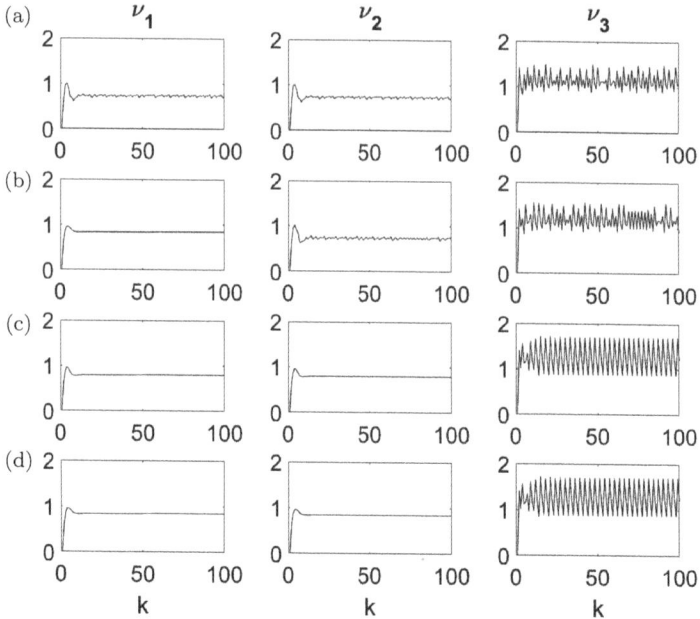

Figure 11.3. Firing-rate sequences corresponding to the circuit modification and segregation cases depicted in Fig. 11.2. Full connectivity (a) results in mutual interference between the firing modes. Neuron 1 isolation by synaptic silencing (b) results in interference-free firing of that neuron. Neuron 3 isolation by synaptic silencing (c) results in interference-free firing, while neurons 1 and 2 fire in interference-free synchrony. Mutual isolation of the three neurons by complete synapse silencing allows each of them to fire in its own characteristic mode (Baram, 2018).

be seen in Figs. 11.3(c) and 11.3(d) that there is a slight difference between the firing modes of neurons 1 and 2 in cases 11.2(c) and 11.2(d). This implies that, while in both cases there is complete synchrony between the two neurons, the 2-neuron circuit connectivity in case (c) does not allow each of the two neurons complete freedom to fire in its own independent firing-rate mode. We call such condition *restrained synchrony*.

It follows that, while the Hebbian paradigm may result in strong inhibitory effects between asynchronously firing neurons and circuits, it need not necessarily completely eliminate mutual interference altogether. On the other hand, the polarity-gated synapse silencing mechanism simply eliminates directional interaction, removing the corresponding interference.

11.5   Firing mode control by synchronous polarity-gated circuit segregation

A cortical circuit of identical neurons may be segregated by polarity-gated synapse silencing into internally synchronous, externally asynchronous subcircuits. Circuit and subcircuit sizes can be further reduced by polarity-gated membrane silencing. As we show next, circuit and subcircuit size modification results in firing-rate mode modification, which may serve different cortical functions, even when the neurons are identical.

Consider first a fully connected circuit of n identical neurons firing in synchrony. As $N \to \infty$, Eqs. (8.6) and (8.7) will take $\omega(k)$ to its constant limit value (Cooper *et al.*, 2004), ω, which, in turn, implies that τ_ω attains an infinitely large value (Baram, 2017a). Equation (8.4) then yields the following firing-rate model for each of the circuit's neurons

$$v(k) = \begin{cases} f_1(v(k-1)) = \lambda_1 v(k-1) \text{ for } \beta n \omega v(k-1) + \beta u \leq 0 \\ f_2(v(k-1)) = \lambda_2 v(k-1) + \beta u \text{ for } \beta n \omega v(k-1) + \beta u > 0 \end{cases}$$

$$(11.6)$$

where $\lambda_1 = \alpha = \exp(-1/\tau_m)$, $\lambda_2 = \alpha + \beta n \omega$ and $\beta = 1 - \alpha$. As explained in Chapter 10, the transition points from one global attractor type of the map to another are defined by the parameters $\lambda_1, \lambda_2, c_1 = 2\lambda_1 \lambda_2 + 1 + \sqrt{1 + 4\lambda_1^2}$ and $c_2 = \lambda_1 \lambda_2 + 1$. Clearly, Eq. (11.6) will produce a different mode of firing-rate dynamics for each circuit size n. This is illustrated by Fig. 11.4, where the four subfigures depict global attractors of circuits having identical neurons, but different primal circuit sizes. It can be seen that different circuit sizes yield different attractor types, even when the individual neuron's parameters, external activation and initial conditions are identical.

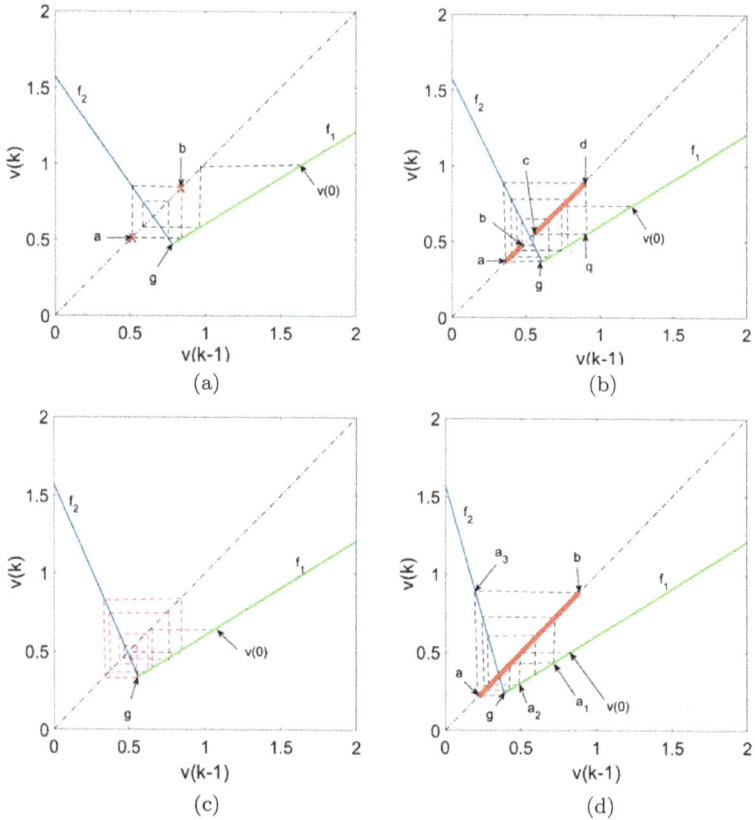

Figure 11.4. Global attractor types corresponding to fully connected synchronous identically parametrized circuits of primal sizes: (a) two neurons (oscillatory), (b) five neurons (largely-oscillatory), (c) seven neurons (fluctuating), and (d) eleven neuron (chaotic). The global attractors are represented by red line segments or by red x (Baram, 2018).

Specifically, the circuits represented by Fig. 11.4 obey the model of Eqs. (8.4)–(8.8). Keeping in mind the linear ratio between membrane potential and firing-rate, 7.2 ± 0.6 spikes $\cdot$ sec$^{-1} \cdot mV^{-1}$ (Carandini and Ferster, 2000), we uniformly assume for illustration purposes the parameter values $u = 4, \tau_m = 2, \tau_\omega = 300, \tau_\theta = 0.1$ in units of mV and mSec, respectively. We further assume zero initial conditions, with $N = 100$ taking $\omega(k)$ to its constant limit ω, and the primal circuit size values specified in the following yielding the corresponding modal parameter and attractor condition values:

(a) $n = 2$, yielding $\omega = -2.5695, \lambda_1 = 0.6065, \lambda_2 = -1.4155, c_1 = 0.8550, c_2 = 0.1415$

(b) $n = 5$, yielding $\omega = -1.3128, \lambda_1 = 0.6065, \lambda_2 = -1.9762, c_1 = 0.1748, c_2 = -0.1987$

(c) $n = 7$, yielding $\omega = -1.0152, \lambda_1 = 0.6065, \lambda_2 = -2.1896, c_1 = -0.0840, c_2 = -0.3280$

(d) $n = 11$, yielding $\omega = -0.8791, \lambda_1 = 0.6065, \lambda_2 = -3.1983, c_1 = -1.3076, c_2 = -0.9398$

An examination of the conditions for the global attractor types specified in Chapter 10 shows that these cases represent (a) oscillatory between points a and b, (b) largely-oscillatory between line segments $[a, b]$ and $[c, d]$, (c) fluctuating between small and large steps and (d) chaotic by the Li-Yorke "period three" (here, $a_1 \to a_2 \to a_3 \to a_1$) criterion (Li and Yorke, 1975) global attractors, as ratified by the cobweb diagrams in Fig. 11.4.

Cobweb diagrams are essentially restricted to scalar evolution and, therefore, the above analysis and, for comparison purposes, the corresponding simulations have considered synchronous circuits of identical neurons under the same initial conditions. However, a broader notion of synchronization under different initial conditions can be illustrated by simulation. For example, case (a) above, representing two fully connected circuit of two identically parametrized neurons under the same initial conditions, is now illustrated in Fig. 11.5 under different initial conditions, $v_1(0) = 1$ and $v_2(0) = 2$, respectively. The simulated sequences of firing-rate corresponding to the two neurons are shown in Fig. 11.5(a). It can be seen that, following different initial periods (Fig. 11.5(b)), the two neurons converge to fully synchronized behaviors (Fig. 11.5(c)). We have further found similar synchronization of fully connected neurons under different initial conditions in a variety of simulated cases.

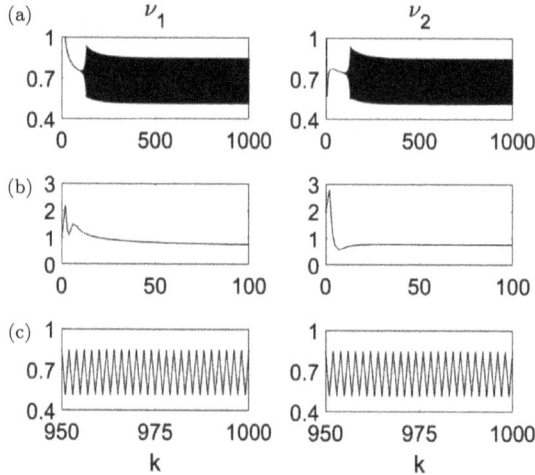

Figure 11.5. Synchronized oscillations in a circuit of two identical neurons under different initial conditions. The two sequences (a) have different initial responses (b), synchronizing in time (c) (Baram, 2018).

11.6 Neural circuit segregation capacity

Neural circuit segregation by inter-neuron synapse silencing implies the division of a circuit into isolated subcircuits. Silent neurons (neurons having membranes with negative polarity) are discounted. Given a fully connected neural circuit of n active neurons (neurons having membranes with positive polarity), consider an isolated sub-circuit of m fully connected neurons. Each of the synapses of the m^2 synapses (including self-synapses) connecting the neurons of the subcircuit to each other can have either positive or negative polarity. There are, then, 2^{m^2} polarity states of the said sub-circuit. As there are

$$\binom{n}{m} = \frac{n!}{(n-m)!m!} \qquad (11.7)$$

possible selections of m out of n neurons, the maximal possible number of segregated subcircuit polarity patterns is

$$P(n) = \sum_{m=0}^{n} \binom{n}{m} 2^{m^2} \qquad (11.8)$$

constituting the segregated polarity code size for subcircuits of m out of n neurons. This number grows very fast with n. For instance, $P(1) = 3$, $P(2) = 21$, $P(3) = 567$, $P(5) = 33,887,403$ and $P(7) = 563,431,696,713,567$.

Following the segregation of a subcircuit of m active neurons out of a fully connected circuit of n active neurons by inter-neuron synapse silencing, we have two externally isolated, internally fully-connected subcircuits, one of m neurons and the other of $n - m$ neurons. Each of these subcircuits has its code of polarities, as specified in Chapter 4. Further inter-neuron synapse silencing can further divide each of the subcircuits into smaller fully-connected subcircuits, etc. Such segregation results in a total polarity code size smaller than that represented by Eq. (11.8). For instance, a circuit of neurons can divide in 3 ways into two isolated subcircuits, one of 2 neurons and one of 1 neuron, producing a total polarity code of size $3 \times 3 \times 21 = 189$ (clearly smaller than 567). When the subcircuits of two neurons further divide into segregated subcircuits of one neuron each, the total polarity code size becomes $3 \times 3 = 9$. While circuit segregation, by virtue of circuit size reduction, clearly reduces the polarity code size of the original circuit, it increases the information representation capacity of the circuit, due to the increased number of subcircuit created. For instance, while a fully connected unsegregated circuit of 3 neurons can represent, at a time, a single polarity word, representing the polarity states of all the membranes and synapses involved, the same 3 neurons can store 2 (albeit shorter) words when segregated into two isolated subcircuits, one of 2 neurons and the other of 1 neuron), or 3 polarity words, when segregated into three isolated subcircuits of 1 neuron each. The maximal information representation capacity of an n- neuron circuit is then, n words, each represented by the polarity state of a single neuron.

11.7   Synaptic and somatic elimination: From pruning to senescence

Permanent synaptic and axonal silencing have been addressed as synapse and axon elimination. The role of synaptic elimination, observed in humans (Huttenlocher, 1979; Huttenlocher *et al.*, 1982;

Huttenlocher and Courten, 1987) and in animals (Eckenhoff and Rakic, 1991; Bourgeois, 1993; Bourgeois and Rakic, 1993; Rakic *et al.*, 1994; Innocenti, 1995) has been generally perceived as the removal or "pruning" of redundant or weak synapses for the improvement of neural circuit performance. Although structural circuit modification has been suggested in general terms as means for long-term memory (Balice-Gordon and Lichtman, 1994), the specific function and mechanization of synapse elimination have remained essentially unclear. Reports that focal blockade of neurotransmission is more effective in synapse elimination than a whole junction blockade (Balice-Gordon and Lichtman, 1994) and that synapse elimination precedes axon dismantling (Balice-Gordon *et al.*, 1993) have been challenged by claims that synapse elimination is a consequence of whole axon removal (Vanderhaeghen and Cheng, 2010). Dynamic firing effects of neural interaction under synapse elimination have been experimentally observed, noting that "active synaptic sites can destabilize inactive synapses in their vicinity" (Balice-Gordon and Lichtman, 1994), although such effects may involve synapse silencing and reactivation (Atwood and Wojtowicz, 1999) rather than synapse elimination. While early studies have associated synapse elimination with early development (Balice-Gordon and Lichtman, 1994; Culican *et al.*, 1998) and childhood (Chechik *et al.*, 1998), others have extended it to puberty (Iglesias *et al.*, 2005) and, depending on brain regions, to age 12 for frontal and parietal lobes, to age 16 for the temporal lobe, and to age 20 for the occipital lobe (Giedd *et al.*, 1999). Yet, Alzheimer's disease (Horn *et al.*, 1996), grey matter (Mechelli *et al.*, 2004) and cognition (Craik and Bialystok, 2006) studies, and persistent evidence of molecular processes involved in synaptic elimination throughout life (Lee *et al.*, 2016) have suggested its relevance all the way to senescence. Connectivity changes due to synapse elimination (Dennis and Yip, 1978; Huttenlocher, 1979) have been suggested as means for long-term memory (Balice-Gordon *et al.*, 1993), and followed by studies of structure (Balice-Gordon and Lichtman, 1994), information capacity and cortical segregation (Baram, 2017b).

11.8 Discussion

Extending the notion of local polarity, experimentally discovered in the separate forms of membrane and synapse silencing and reactivation about two decades ago, to the unified notion of polarity-gated neural circuits, we have shown that circuit polarization segregates a neural circuit into interference-free, internally-synchronous, externally-asynchronous subcircuits. We have further shown that chaotic, oscillatory, fixed-point and silent firing-rate modes are governed by mixed, positive and negative polarity gating, respectively, maintaining the firing variety of the different neurons and allowing each of the neurons its unique expression by firing-rate dynamics.

Synchronous circuits can be further segregated into internally synchronous subcircuits of same or different sizes, that can simultaneously perform different cortical functions. Internally synchronous circuits of different sizes will operate in different firing-rate modes, which will, in turn, segregate the different functions performed. In certain contexts, such as learning and memory discussed in the following chapter, different firing-rate modes would define the language of the dynamics generated, stored, retrieved and conveyed. The cortical ability to generate, by neural circuit polarization, a large variety of firing-rate modes guarantees a combinatorial richness of both function and information repertoires. Yet, as illustrated by Fig. 11.4, circuits of primal size greater than 7 yield (specifically 11 or more) yield chaotic behavior. This may be viewed as another evidence of "the magical number seven" (Miller, 1956), limiting working memory capacity (Pribram *et al.*, 1960). Finally, we note that while synaptic and somatic elimination may be regarded as the long-term realization of the corresponding polarization, they have far-reaching age-related consequences not only in terms of lasting effects, such as long-term memory, but also in terms of senescence and impairment.

Chapter 12

Learning and Memory by Circuit Polarization

12.1 Introduction

Given neuronal parameters, specifically, the time constants τ_m, τ_ω and τ_θ, the only variable internal properties in the spike response and plasticity models (e.g., Eqs. (2.15)–(2.18)) are the synaptic weights. These, under certain conditions (Rosenblatt, 1958; Oja, 1982; Cooper *et al.*, 2004; Castellani, 2017), are subject to convergence to constant limit values. Such convergence, leading to memory, may be referred to as *learning*. While invariant synaptic weights or their elimination represent *long-term* memory, persistent changes in the dynamics of the synaptic weights and changes in the dynamic modes of firing-rates, which fall in the category of generalized *metaplasticity* (Baram, 2017a), define another form of memory, which may be termed *short-term* memory. Short-term memory periods are marked by transitions from one firing-rate mode to another, brought about by changes in circuit polarity. Exertion of the same external inputs that have instigated the original learning, or memory, will reproduce the same circuit behavior, represented by firing-rate modes. Moreover, as illustrated in Fig. 11.5, the memorized firing-rate attractors are robustly retrieved regardless of different initial conditions. Suppose that the operand of the function f in Eq. (8.4) becomes negative and stays negative for some time. This may be a consequence of a drop in the level of the external input potential, $I_i(k)$, or by an unforced change in membrane potential resulting from slow dissipation of

electric charge. As explained in Chapter 11, and demonstrated by
Eq. (11.1) and Fig. 11.1(a), this will result in neuronal silencing,
yielding partial circuit memory loss and, as all external inputs to
the circuit drop to sufficiently low levels, a total loss of circuit
activity. This may be regarded as memory concealment, but not
as loss of memory, as the convergent values of the synaptic weights,
representing memory, are still in effect. Memory restoration will occur
when the circuit polarity is the same as the one which was valid when
the original memory was instigated, and when the external inputs
valid at that time are recovered. Since, as shown in Chapter 9, firing-
rate modes are uniquely determined by internal neuron properties
and by circuit polarity, the firing-rate modes are uniquely recovered
by restoration of the corresponding circuit polarity pattern. Such
memory restoration may be partial, affecting, at different times,
individual neurons or subcircuits. The respective roles of circuit
polarity and external input levels in memory concealment, restoration and modification are explained and demonstrated in the sequel.
As noted in Section 10.5, the difference between the formation of a
global attractor, and the outcome of such formation, constitutes a
distinction between learning and memory in the context of a firing-
rate dynamics-based theory of metaplasticity. In addition to such
general notions, addressed in Sections 12.2–12.4, this chapter also
addresses the notion of "associative memory", which, much like the
notion of "working memory" addressed in Chapter 5, and in contrast
to previously published conceptions (e.g., Hinton *et al.*, 1986) is put
in the underlying context of this book, specifically, neural circuit
polarity.

12.2   Short and long-term memory in the individual neuron

In order to demonstrate the basic concepts of short and long-term
memory, we first consider an individual neuron in the following four
cases:

(a) $u = 10$, $\tau_m = 2$, $\tau_w = 10000$, $\tau_\theta = 0.1$, yielding $\lambda_1 = 0.6055$,
$\lambda_2 = -4.7336$, $c_1 = -3.1701$, $c_2 = -1.8711$, $\omega = -13.5720$

(b) $u = 10$, $\tau_m = 2$, $\tau_w = 10000$, $\tau_\theta = 1$, yielding $\lambda_1 = 0.6055$, $\lambda_2 = -1.8160$, $c_1 = 0.3692$, $c_2 = -0.1015$, $w = -6.1569$

(c) $u = 1$, $\tau_m = 2$, $\tau_w = 5$, $\tau_\theta = 1$, yielding $\lambda_1 = 0.6065$, $\lambda_2 = 0.5308$, $w = -0.1925$

(d) $u = -1$, $\tau_m = 2$, yielding $\lambda_1 = \lambda_2 = 0.6065$, $w = 0$

with the notation defined in Chapters 8–11. For each of the cases, the steady-state value of the synaptic weight, w, was calculated by driving Eqs. (8.4)–(8.8), with $n = 1$, $w(0) = 0$ and $v(0) = 1$, to convergence (practically, this was achieved for $N = 100$), and the corresponding values of λ_1, λ_2, c_1 and c_2 were calculated by Eqs. (8.23), (8.24), (10.7) and (10.8), respectively. As can be verified, the conditions stated earlier for the attractor types (a) chaotic, (b) largely-oscillatory, (c) fixed-point and (d) silent are satisfied, respectively, by the parameter values obtained. The four cases are illustrated by cobweb diagrams in Fig. 12.1.

Simulating each of the four cases, as shown in Fig. 12.2, the time domain for the firing-rate $v(k)$, $k = 0, \ldots, 30,000$ is divided into three subdomains. In the first subdomain, $k = 0, \ldots, 10,000$, we observe convergence of the feedback synaptic weight w, and a simultaneous convergence of the firing-rate to a global attractor. Specifically, while in cases (a), (b) and (c) the positive centered external activation u enforces convergence of w to a negative limit value and to persistent firing-rate modes, the negative u in case (d) results in membrane silencing, as predicted by Eq. (11.1) and illustrated by Fig. 11.1(a). Membrane silence may last indefinitely, representing loss of memory, hence the terminology *short-term memory*. In the second subdomain, $k = 10,000 \ldots 20,000$, membrane silencing and, consequently, memory concealment, was accomplished by changing u from 10 to -10 in cases (a), (b) and (c), and from -1 to -10 in case (d). Membrane silence in cases (a), (b) and (c) was interrupted in the subdomain $k = 20,000 \ldots 30,000$ by reinstating the original levels of activation, which results in restoration of the memorized (or learned) mode of firing-rate, constituting *long-term memory*.

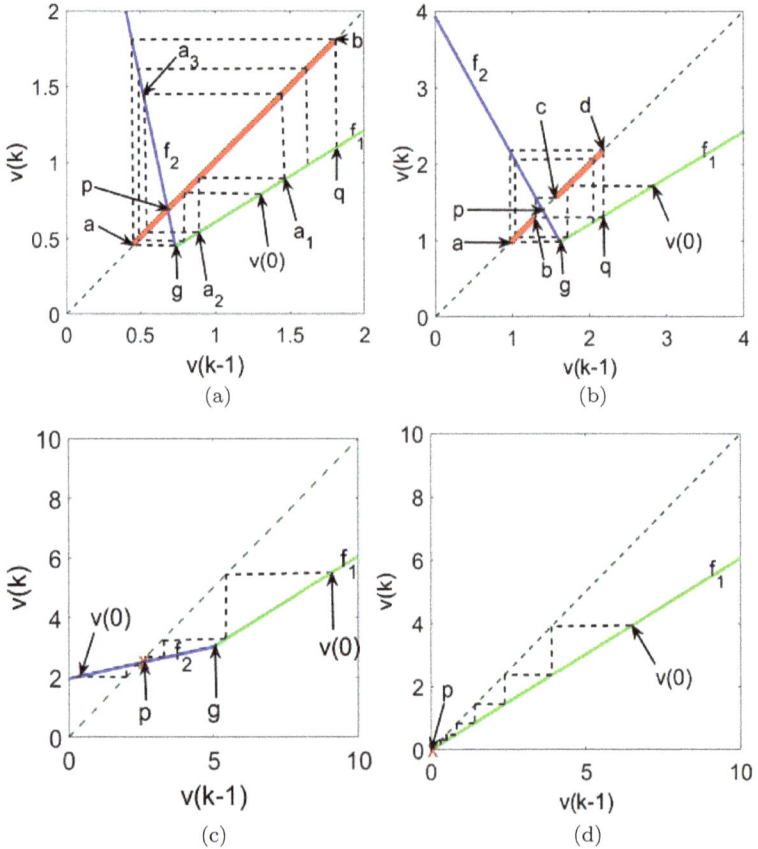

Figure 12.1. Global attractor code of firing-rate dynamics in individual neuron with active self-synapse and silenced synapses of circuit pre-neurons: (a) chaotic, (b) largely-oscillatory, (c) fixed-point and (d) silent. Cobweb trajectories are represented by black dashed lines and converge to global attractors represented by red line segments or red x. The model is specified by Eqs. (8.4)–(8.8) for $n = 1$, and the parameter values for each of the global attractors are specified in the text (Baram, 2018).

12.3 Short and long-term memory in synchronous circuit

As explained and demonstrated in Chapter 11, neural circuits can be segregated by synapse silencing into synchronous subcircuits. External input decline, or membrane potential decline by spontaneous

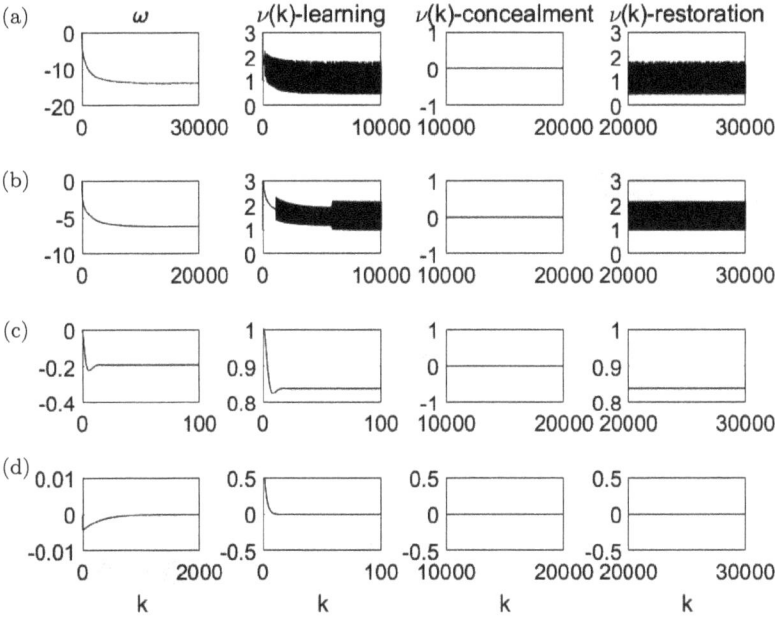

Figure 12.2. Learning (first and second columns), short-term memory (second column), memory concealment (third column) and long-term memory restoration (fourth column) in the individual neuron are represented by firing-rate modes in the four cases considered (Baram, 2018).

electric discharge, will result in individual neuron silencing. A change in the number of active circuit neurons will result in changing firing-rate dynamics of the remaining circuit neurons, possibly to the point of total circuit silence and memory concealment. Restoration of external activation will result in partial, or complete, return of circuit activity to its previous dynamics.

In order to demonstrate memory deterioration, concealment and restoration, representing short and long-term memory in a synchronous circuit, consider a circuit of three neurons, all having the same parameter values $\tau_w = 300$, $\tau_\theta = 0.1$, $\tau_m = 2$ and $u = 4$. In time, the circuit undergoes several different stages of individual neuron silencing and reactivation, as depicted in Fig. 12.3. Figure 12.3(a) describes initial learning and memory, followed by memory deterioration due to a sequence of neuronal silencing,

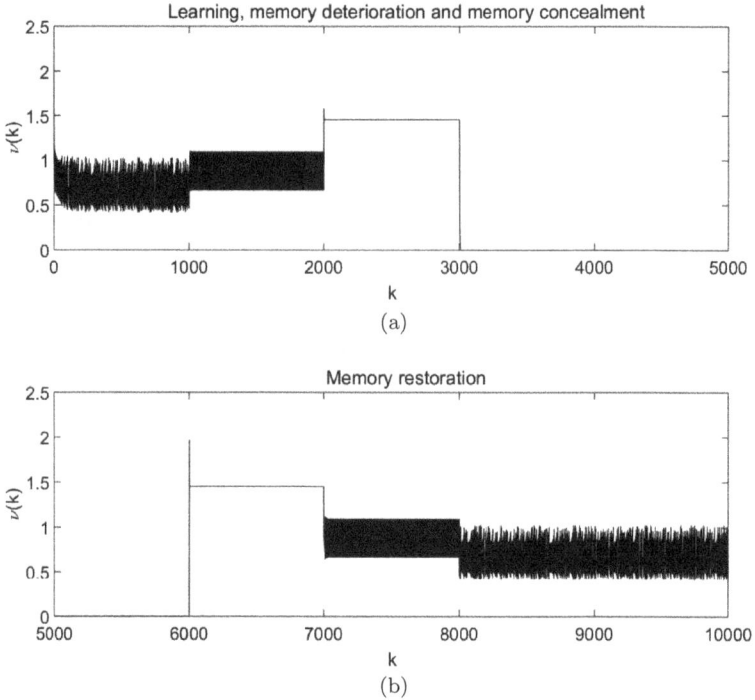

Figure 12.3. Learning, memory deterioration and concealment (a) and memory restoration (b) represent short and long-term memory, respectively, in synchronous circuit. A 3-neuron synchronous circuit undergoes a sequence of learning, converging to a chaotic firing-rate mode ($k = 0, \ldots, 1000$), followed by 1 neuron silencing yielding an oscillatory firing-rate mode of the active 2-neuron subcircuit ($k = 1001, \ldots, 2000$), a second neuron silencing yielding a constant firing-rate mode of the remaining active neuron ($k = 2001, \ldots, 3000$), and a third neuron silencing yielding complete memory concealment ($k = 3001, \ldots, 5000$). Neuronal activation in reversed order ($k = 5001, \ldots, 10000$) ends in restoration of the learned chaotic firing-rate mode (Baram, 2018).

achieved by changing the value of the corresponding centralized membrane activation from $u = 4$ to $u = -1$. During the time interval between $k = 0$ and $k = 1000$, the circuit experiences learning, by the end of which it reaches a chaotic mode of firing-rate. At $k = 1001$ one of the neurons becomes silent. The remaining active part of the circuit, constituting a synchronous subcircuit of two neurons, displays an oscillatory mode of firing-rate (identical to case (a) in

Fig. 11.4, having the same circuit size of 2 and the same parameters), until $k = 2000$. At this point another neuron becomes silent, and, following an abrupt response to the change in circuit size at $k = 2001$, the remaining active neuron displays a constant (fixed-point) mode of firing-rate. This last active neuron becomes silent at $k = 3001$, and the entire circuit remains silent until $k = 5000$. The sequence of neuron silencing thus described represents memory deterioration, ending in complete memory concealment.

The memory deterioration process described by Fig. 12.3(a) is reversed in Fig. 12.3(b) by reactivating, in sequence, each of the neurons by applying $u = 4$, as in the original learning process. The circuit remains silent until, at $k = 6001$, one of the neurons becomes active again, producing a constant firing-rate. At $k = 7001$ a second neuron becomes active, and the subcircuit of two active neurons produces an oscillatory mode of firing-rate. At $k = 8001$, the third neuron becomes active as well, and the original external activation of all three neurons is restored. It can be seen that the circuit activity is back to the original chaotic mode.

12.4 Memory modification

Memory modification may result from changes in neuronal circuit polarity and in membrane activation level. While the circuit polarity pattern uniquely determines which neurons are active, the dynamic nature of the firing-rate produced by a neuron as a result of restored activity will depend on the level of external activation as well. As can be seen in Eq. (8.22) for $n = 1$, the actual value of u determines the intersection point βu of the function $f_2(v(k-1))$ with the axis $v(k)$, but not the slope λ_2 of $f_2(v(k-1))$, defined by Eq. (8.24). Therefore, the value of u affects the amplitude, but not the characteristic firing-rate mode of the individual neuron (this can be further ratified by examination of the cobweb diagrams in Fig. 12.1).

Repeating the simulations of the individual neuron depicted in Fig. 12.2, with the retrieval activation values changed from $u = 10$, used in the learning stage of cases (a)–(c), to (a) $u = 1$, (b) $u = 0.1$, (c) $u = 0.01$ and (d) $u = -1$ applied in retrieval, we obtain the

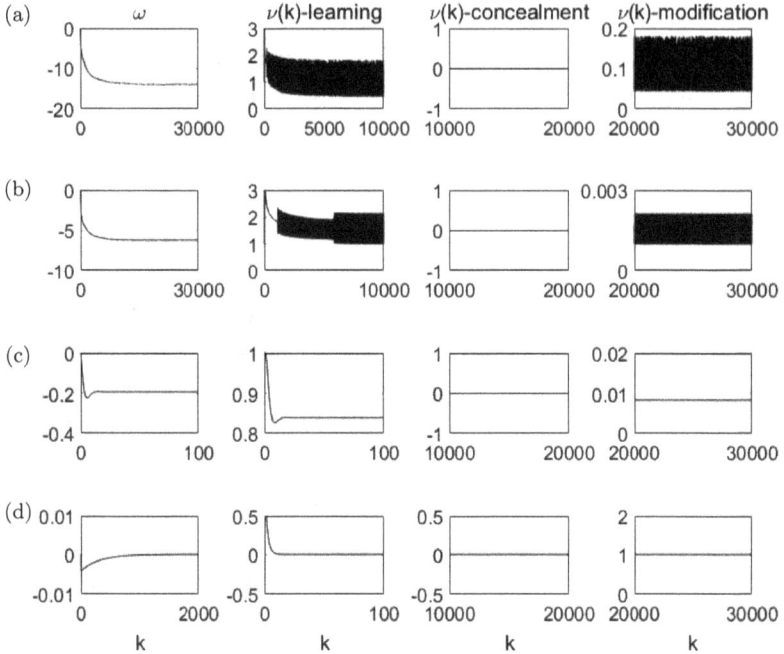

Figure 12.4. Using different activation values in learning (second column) and retrieval (fourth column) by an individual neuron (cases (a)–(d) of Fig. 12.1) results in different firing-rate amplitudes (Baram, 2018).

results depicted in Fig. 12.4. It can be seen that while the evolution of the feedback synaptic weight and the firing-rates during learning depicted in Fig. 12.4 are identical to those depicted in Fig. 12.2, the firing-rate sequences in retrieval have different amplitudes in the two figures.

In a synchronous neural circuit, a process of memory deterioration or restoration may stop at any stage, producing what might be considered a "partial memory", an "illusion", or an "innovation". For instance, if the process of memory deterioration in the synchronous 3-neuron circuit considered in Section 12.3 stops between $k = 1001$ and $k = 2000$, then memory will continue as an oscillatory sequence produced by a 2-neuron circuit, which is different from the original chaotic sequence. Similarly, if the process of deterioration stops

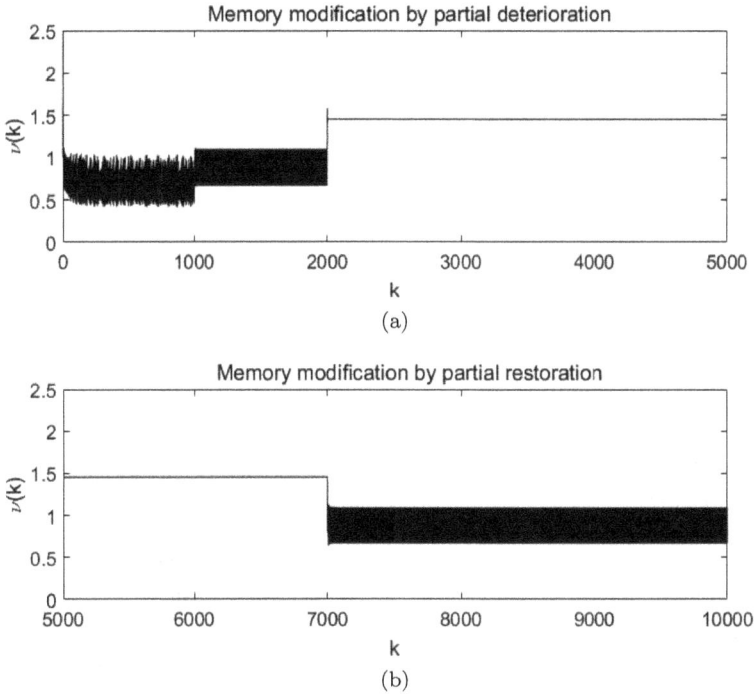

Figure 12.5. Memory modification in a synchronous 3-neuron circuit. (a) Following learning, producing chaotic firing-rate ($k = 0, \ldots, 1000$), silencing of one neuron ($k = 1001, \ldots, 2000$), followed by silencing of another neuron ($k = 2001, \ldots, 7000$) takes the circuit to oscillatory, then to constant firing-rate, contrasting the original chaotic mode. (b) Reactivating one of the silenced neurons takes the circuit back to oscillatory firing-rate, again contrasting the original chaotic mode (Baram, 2018).

between $k = 2001$ and $k = 3000$, then memory will continue as a constant sequence produced by a single neuron, as illustrated in Fig. 12.5(a).

The same is true for the memory restoration sequence, which, if stopped between $k = 6001$ and $k = 7000$, will produce memory of a constant sequence produced by a single neuron, while, if stopped between $k = 7001$ and $k = 8000$, will produce memory of an oscillatory synchronous sequence in a 2-neuron circuit, as illustrated in Fig. 12.5(b).

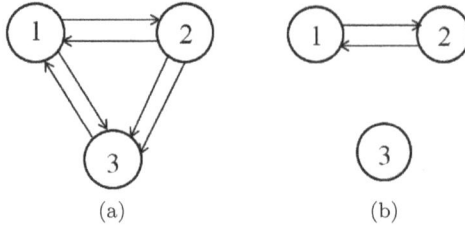

(a)   (b)

Figure 12.6. Associative and segregated memory. (a) Three-neuron circuit associating a synchronous subcircuit of two neurons (neurons 1 and 2) with a single neuron (neuron 3) having different parameters. (b) Two segregated circuits, one of two synchronous neurons (1 and 2), the other of one neuron (neuron 3) associating by simultaneous, yet asynchronous firing.

12.5   Associative memory by circuit polarization and segregation

When different memories are stored and concealed in the neurons of a neural circuit, associations between them are defined by circuit connectivity. Memory will be retrieved in an associative manner when neurons and subcircuits storing different memories are activated simultaneously. Mutual interference may be eliminated by inter-synapse silencing. Consider the two circuits displayed by Fig. 12.6(a) and 12.6(b). The three neurons, marked 1, 2 and 3, appearing in both circuits, have the following parameters, indexed accordingly:

$$u_1 = u_2 = 1, u_3 = 4, \ \tau_{m,1} = \tau_{m,2} = 1, \ \tau_{m,3} = 2, \ \tau_{w,1} = \tau_{w,2} = 5,$$

$$\tau_{w,3} = 100, \ \tau_{\theta,1} = \tau_{\theta,2} = 1, \ \tau_{\theta,3} = 1$$

Figure 12.7 shows, progressively in time, the simulated firing-rates of the three neurons $v_1(k)$, $v_2(k)$ and $v_3(k)$, respectively. The first column ("learning"), representing the first 100 time samples, shows convergence of the firing-rate modes of the three neurons connected in the associative arrangement of Fig. 12.6(a). It can be seen that, while neurons 1 and 2 fire in concert, neuron 3 has its own firing-rate mode, as might be expected from its different parameters. While each of the subcircuits conveys a distinctly different type of information (fixed point vs oscillatory firing-rate mode), the

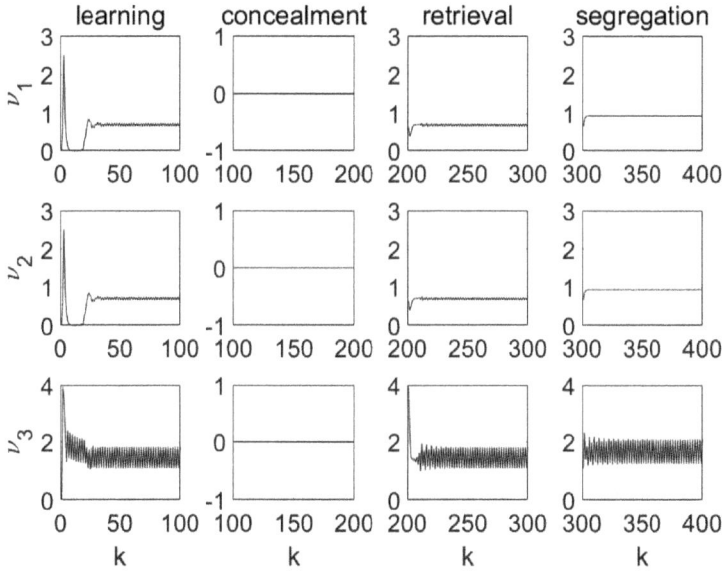

Figure 12.7. Associative and segregated memory in asynchronous 3-neuron circuit. Firing-rates of neurons 1, 2 and 3 in the circuits displayed by Fig. 12.6 are shown in the three rows. The first three columns represent learning, concealment and retrieval stages of associative memory corresponding to the circuit displayed in Fig. 12.6(a), while the last column represents interference-free firing-rates of the segregated circuit displayed in Fig. 12.6(b).

firing-rate modes of the two subcircuits (neurons 1 and 2 vs neuron 3), while clearly interfering with each other, can be said to be *associated* by the circuit. The second column of Fig. 12.7 ("concealment"), corresponding to the next 100 time samples, represents membrane silencing of the three neurons, during which none of the information stored is displayed. The third column of Fig. 12.7 ("retrieval") displays the firing-rates aroused by reactivating the membranes of the three neurons. It reproduces the three original characteristic firing-rate modes (first column, with, however, a considerably shorter initial transient stage, as the synaptic weights have already reached their limit values, hence, memory, associated with the learning stage). The fourth column of Fig. 12.7 ("segregation") corresponds to the segregated circuit displayed in Fig. 12.6(b). It can be seen that with the corresponding synapses silenced, each of the two subcircuits, the first consisting of neurons 1 and 2 and the second consisting of neuron 3, fires in its characteristic firing mode, without the interference seen in the original interaction (first column). Yet, the combined action of the three-neuron circuit may be regarded as associative memory, in the sense that the two subcircuits are activated at the same time, evoking memories which, in the past were associated by simultaneous activity. Similarly, firing modes of different subcircuits, "learned" at different times, may be recreated, hence, associated, at the same time.

12.6   Discussion

As plasticity is a fundamental property of the neuron, it seems conceivable that all cortical functions involve memory, manifested by recreation of a previous neural firing experience. Yet, as we have demonstrated, the circuit-wide expression of evoked memory can take on different forms, associated with circuit connectivity. The latter is controlled by circuit polarity, which not only segregates and reconnects circuits and subcircuits, but also determines their firing-rate modes. While we have shown a variety of learning and retrieval scenarios by individual neurons, by synchronous circuits, and by associative and segregated asynchronous circuits, the wealth of information that can be stored, retrieved and manipulated by circuit and subcircuit segregation and reconnection is combinatorial in the number of neurons involved.

Part III
Cortical Quantum Effects

Chapter 13

Some Quantum Computation Preliminaries

13.1 Introduction

Just as the phenomena of quantum physics do not appear to be in natural coherence with those of classical physics, the mathematics of quantum computation requires a certain departure from classical mathematics. Neither seems immediately intuitive. Even the basic definitions are far from being self-explanatory. It is therefore rather hopeless to try to follow intuitive notion. Instead, the underlying set of basic definitions should be familiarized in a completely formal and accurate manner. Even this step requires considerable training. The following preliminaries represent a bold, yet modest attempt to present some basic definitions, so as to set the stage for the introduction of some cortically-based concepts, such as associative memory, in a quantum theoretic context in the next chapter. While the potential advantages of this context in terms of storage capacity, retrieval accuracy and time complexity will be clearly specified, some surprising similarity to classical findings, such as the linguistic plausibility of small neural circuits under subcritical connectivity probability of random graphs will be noted as well. Yet, staying within the realm of classical mathematics and avoiding these two chapters would seem to represent an understandable reader's choice.

13.2 States and qubits

The basic entity of classical computation is the classical bit. Each classical bit can have one of two values, 0 and 1. The state of any finite physical system that can be found in a finite number of states can be described by a string of bits. A string of n bits represents one of 2^n possible states of a system enumerated $0, \ldots, 2^n - 1$. In quantum computation, the basic entity is called a qubit (quantum bit). The qubit can have the analogue values $|0\rangle = [1\ 0]$ and $|1\rangle = [0\ 1]$, known as the computational basis states, where $|\cdot\rangle$ is the Dirac notation. Qubit basis states can also be combined to form product basis states. For example, two qubits could be represented in a four-dimensional linear vector space spanned by the following tensor product basis states: $|00\rangle = [1000]$, $|01\rangle = [0100]$, $|10\rangle = [0010]$, and $|11\rangle = [0001]$.

Yet, the qubit can also have any other value that is a linear combination of $|0\rangle$ and $|1\rangle$:

$$\Psi = \alpha|0\rangle + \beta|1\rangle \tag{13.1}$$

where α and β are any complex numbers (called the amplitudes of the basis states 0 and 1, respectively), such that $|\alpha|^2 + |\beta|^2 = 1$. Consequently, the qubit can be in any one of an infinite number of states described by unit vectors in a two-dimensional complex vector space. The unary representation of a qubit can be given as a vector of two values

$$|\psi\rangle = \begin{pmatrix} \alpha \\ \beta \end{pmatrix} \tag{13.2}$$

Analogously to the classical bit string, qubit strings describe the state of a system. A two qubit system comprising two qubits $|a\rangle = \alpha|0\rangle + \beta|1\rangle$ and $|b\rangle = \gamma|0\rangle + \delta|1\rangle$ is described by the tensor product of the two qubits $|a, b\rangle \equiv |a\rangle \otimes |b\rangle = \alpha\gamma|00\rangle + \alpha\delta|01\rangle + \beta\gamma|10\rangle + \beta\delta|11\rangle$.

13.3   Measurement

The measurement of a qubit reveals only one of two possible outcomes. The value of α and β cannot be extracted from the measurement of a qubit. Instead, when measuring the qubit $\alpha|0\rangle + \beta|1\rangle$ in the computational basis, the result can be either 0 or 1 with probabilities $|\alpha^2|$ and $|\beta|^2$ respectively. For example, the state $\left(\frac{1}{\sqrt{2}}|0\rangle + \frac{1}{\sqrt{2}}|1\rangle\right)$, when measured, yields any of the two results 0 or

1 with probability $\frac{1}{2}$. The measurement operation is not reversible and, once made, the qubit no longer exists in its state before the measurement. Measurements can be performed in different bases. For example, measuring the qubit $\alpha|0\rangle + \beta|1\rangle$ in the Hadamard basis defined by the two basis states $|+\rangle \equiv (\frac{1}{\sqrt{2}}|0\rangle + \frac{1}{\sqrt{2}}|1\rangle)$ and $|-\rangle \equiv (\frac{1}{\sqrt{2}}|0\rangle - \frac{1}{\sqrt{2}}|1\rangle)$ gives $|+\rangle$ with probability $\frac{(\alpha+\beta)^2}{2}$ and $|-\rangle$ with probability $\frac{(\alpha-\beta)^2}{2}$, since $\alpha|0\rangle + \beta|1\rangle = \frac{\alpha+\beta}{\sqrt{2}}|+\rangle + \frac{\alpha-\beta}{\sqrt{2}}|-\rangle$.

An n-qubit system can be either measured completely or partially. When measured partially, the unmeasured subsystem can retain quantum superposition and further quantum manipulations can be performed upon it. However, any measurement can be delayed to the end of the computation process.

13.4 Operators and oracles

In quantum computation, a system changes its state under a unitary quantum operator U from $|\Psi\rangle$ to $U|\Psi\rangle$. An operator U can be described as a $2^n \times 2^n$ matrix operating on the unary representation of the system state. A unitary operator satisfies $UU^\dagger = I$, where $U^\dagger$ is the conjugate transpose of U (transpose the matrix U then substitute the conjugate complex of each element in the matrix). Quantum operators can be implemented using quantum gates, which are the analogue of the classical gates that compose classical electrical circuits. In this analogy, the wires of a circuit carry the information on the system's state, while the quantum gates manipulate their contents to different states. For example, the Hadamard operator

$$H = \frac{1}{\sqrt{2}}\begin{pmatrix} 1 & 1 \\ 1 & -1 \end{pmatrix} \tag{13.3}$$

transforms a qubit in the state $|0\rangle$ into the state $|+\rangle$ and the state $|1\rangle$ into the state $|-\rangle$. The Hadamard operator can also be seen as operating on n-qubits by the tensor product of single qubit Hadamard operators. Each qubit is then transformed according to the single qubit Hadamard transform, that is

$$H^{\otimes n}|x_{n-1}, \ldots, x_1, x_0\rangle = H|x_{n-1}\rangle \otimes \ldots \otimes H|x_1\rangle \otimes H|x_0\rangle \quad (13.4)$$

An oracle is a "black box" which, by answering a query, produces an input to another algorithm. Given a function f, a quantum oracle U_f is a reversible oracle that accepts a superposition of inputs $|x, y\rangle$ and produces a superposition of outputs according to

$$|x, y\rangle \to U_f|x, y\rangle = |x, f(x) \oplus y\rangle \quad (13.5)$$

as depicted in Fig. 13.1. When the additional qubit is initialized by $|-\rangle$, the oracle is called a quantum phase oracle that gives $f(x)$ in the phase of the state $|x\rangle$ as follows:

$$|x\rangle|-\rangle \to (-1)^{f(x)}|x\rangle|-\rangle.$$

A quantum circuit can process many inputs simultaneously and receive all the outcomes at the output. Consider the case where $|y\rangle = |0\rangle$. Applying the n-qubit Hadamard operator to the $|0\rangle$ state, $|x\rangle = H^{\otimes n}|0\rangle^{\otimes n}$, yields a superposition of all basis states. The resulting output would be $\frac{1}{\sqrt{2^n}} \sum_{i=0}^{2^n} |i\rangle|f(i)\rangle$, which can be calculated as

$$U_f^{\otimes n+1}((H^{\otimes n} \otimes I)|0\rangle^{\otimes n}|0\rangle) = U_f^{\otimes n+1} \frac{1}{\sqrt{2^n}} \sum_{i=0}^{2^n} |i\rangle|0\rangle$$

$$= \frac{1}{\sqrt{2^n}} \sum_{i=0}^{2^n} |i\rangle|f(i)\rangle \quad (13.6)$$

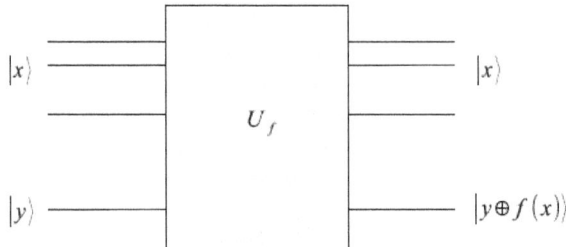

Figure 13.1. A quantum oracle.

13.5 Grover's quantum search algorithm

Given a database of $N = 2^n$ unsorted elements of n bits each, any classical search would require $O(N)$ queries to find a desired element. In 1996, Grover presented a quantum computational algorithm that searches an unsorted database with $O(\sqrt{N})$ operations (Grover, 1996; Boyer *et al.*, 1996). The algorithm performs a series of $O(\sqrt{N})$ unitary operations on the superposition of all basis states that amplify the solution states causing the probability of measuring one of the solutions at the end of the computation to be close to 1. Suppose that the search problem has a set X of r solutions and that we own an oracle function f_X that identifies the solution $x \in X$ according to the following:

$$f_X = \begin{cases} 1, & x \in X \\ 0, & x \notin X \end{cases} \tag{13.7}$$

Any classical algorithm that attempts to find the solution clearly needs to query the oracle N times in the worst case. Grover's algorithm finds the solution with the help of the oracle by querying it only $O(\sqrt{N})$ times.

The quantum phase oracle of the function f_X flips (rotates by π) the amplitude of the states of X, while leaving all other states unchanged. This is done by the operator $I_X = I - 2\sum_{x \in X} |x\rangle$. In matrix formulation, I_X is similar to the identity matrix I except it has -1 as the x elements of the diagonal.

Grover's algorithm starts with the superposition of all basis states created by applying the Hadamard operator to the zero state, $H^{\otimes n}|0^{\otimes n}\rangle$ (shortened by $H|0\rangle$), and goes about performing multiple iterations, in which each iteration consists of applying the phase oracle followed by the operator HI_0H, where I_0 flips the phase of the state $|0\rangle^{\otimes n}$. Grover's iterator is thus defined as

$$Q = -HI_0HI_X \tag{13.8}$$

where the sign "$-$" stands for the global phase flip that has no physical meaning and is performed only for analytical convenience.

The operator in Eq. (13.8) can be viewed in the space defined by the two basis states

$$|\ell_1\rangle = \frac{1}{\sqrt{r}} \sum_{i \in X} |i\rangle \tag{13.9}$$

and

$$|\ell_2\rangle = \frac{1}{\sqrt{N-r}} \sum_{i \notin X} |i\rangle \tag{13.10}$$

as the rotation

$$R = \begin{pmatrix} 1 - \frac{2r}{N} & 2\frac{\sqrt{r(N-r)}}{N} \\ -2\frac{\sqrt{r(N-r)}}{N} & 1 - \frac{2r}{N} \end{pmatrix} \tag{13.11}$$

which is depicted in Fig. 13.2, where the rotation angle is

$$w = \arccos\left(1 - \frac{2r}{N-r}\right) \tag{13.12}$$

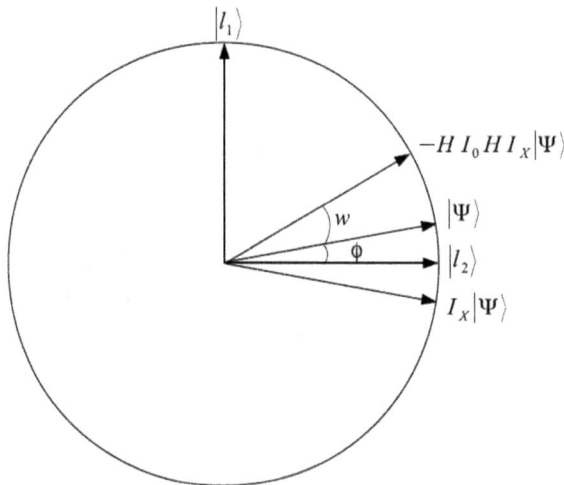

Figure 13.2. The effect of Grover's rotation on the state $|\Psi\rangle$.

The initial state has an angle

$$\phi = \arctan\left(\sqrt{\frac{r}{N-r}}\right) \tag{13.13}$$

with respect to $|\ell_2\rangle$, and after some analysis one can find that applying the operator T times starting from the initial state yields a solution state with a maximal probability that is very close to 1 upon measurement when

$$T = \left\lfloor \frac{\pi}{4}\sqrt{\frac{N}{r}} \right\rfloor = O\left(\sqrt{\frac{N}{r}}\right) \tag{13.14}$$

The algorithm performs $O(\sqrt{N/r})$ iterations, where $r = |X|$ is the number of marked states. Additional improvements were made by Boyer *et al.* (1996) and Brassard *et al.* (1998) coping with an unknown number of marked states in complexity $O(\sqrt{N/r})$.

Biham *et al.* (1999) introduced a third improvement that outputs a marked state when initiated with a state of an arbitrary amplitude distribution.

13.6 Discussion

Quantum computation has been formally found to offer substantial advantages, at least under certain tasks, over classical computation. Specifically, Grover's quantum search algorithm constitutes a sound application of quantum computation to a significant real problem, unequivocally defeating a classical computation approach. Yet, beyond the theoretical support of such application, its practical implementation is far from being efficiently realized. The efforts towards the development of quantum computers, carried out by major companies and research institutes, have been relentless, yet, slow in progress. The difficulties are largely of technological nature. The generation of even a single quantum bit requires enormous effort and perseverance. Yet, even the minutest sign of success has been a major source of encouragement for continued effort. At the same time, the theoretical potential for computational advantages has been driving research towards finding possible roles and neurophysiological implementations of quantum computation for cortical purposes. The following chapter presents the findings of such research.

Chapter 14

Associative Memory by Quantum Set Intersection: The Edge of Small Neural Circuits

14.1 Introduction

The Hebbian memory paradigm was addressed in previous chapters employing so-called classical mathematics. Molecular and neurophysiological manifestations of cortical quantum mechanisms have been suggested (e.g., Ivancevic and Ivancevic, 2010; Clark, 2014). Specifically, Ca^{2+} waves have been hypothesized to drive the Grover's quantum iterator (Eq. (13.8)) while the corresponding oracle (Eq. (13.5)) is regulated by the trisphosphate receptor (IP_3R) channel (Clark, 2014). Employing the notion of quantum set intersection, it has been shown (Salman and Baram, 2012) that the Grover algorithm offers not only considerable reduction in time complexity of associative memory, but an exponential storage capacity exceeding that achieved by earlier propositions (Ventura and Martinez, 2000; Ezhov *et al.*, 2000; Howell *et al.*, 2000; Arima *et al.* 2008; Arima *et al.*, 2009; Miyajima *et al.*, 2010). Here, we emphasize that the quantum approach to associative memory not only improves time and memory capacities, but also resolves the linguistic issue associated with large neural networks, advocating, much like the probabilistic subcriticality addressed in Chapter 4, small neural circuit segregation.

14.2   Quantum set intersection

Quantum set intersection has been shown to provide a general plat-
form for associative memory through pattern equilibrium, pattern
completion and pattern correction (Salman and Baram, 2012).
Consider two sets of marked states, K of size k and M of size m,
and two corresponding oracles f_K and f_M satisfying

$$f_K(x) = \begin{cases} 1, & x \in K \\ 0, & x \notin K \end{cases} \tag{14.1}$$

and

$$f_M(x) = \begin{cases} 1, & x \in M \\ 0, & x \notin M \end{cases} \tag{14.2}$$

with phase oracles I_K and I_M, respectively. The intersection set of K
and M can be found by the following alternating Grover algorithm:

Iterative intersection set search procedure

Given phase oracles I_K, I_M

Denote $Q_K \equiv -HI_0HI_K, Q_M \equiv -HI_0HI_M$

Let $|\Psi\rangle = H|0\rangle^{\otimes n}$

Repeat $|\Psi\rangle = Q_K|\Psi\rangle$, $|\Psi\rangle = Q_M|\Psi\rangle$, $T = O\left(\sqrt{\frac{N}{r}}\right)$ times

Measure $|\Psi\rangle$

It can be shown (Salman and Baram, 2012), that, following
$O(\sqrt{N/r})$ iterations, the above iterative intersection search procedure finds members of the intersection set with probability which is
asymptotically close to 1. The above search procedure assumes that
the size of the intersection set, r, is known in order to determine the
number of iterations. In the more general case where r is unknown
a modification for an unknown number of marked states can be
employed (Boyer *et al.*, 1996).

The following result (Salman and Baram, 2012, Theorem 1) yields
an upper bound on the probability of measuring a state in the
intersection $K \cap M$.

Let I_K and I_M be phase oracles that mark two sets of n-qubit states K and M with $|K|$, $|M| \ll N$. Let us denote $|K| = k$, $|M| = m$, $|K \cap M| = r$,

$$Q \equiv Q_M Q_K \equiv (H I_0 H I_M H I_0 H I_K) \tag{14.3}$$

and $|\Psi(t)\rangle = Q^T H |0\rangle|$. Then, the maximal probability of measuring a state in the intersection $K \cap M$ is achieved at

$$T = \arg\max_i \sum_{x \in K \cap M} |\langle x|\Psi(t)\rangle|^2 = \left\lceil \frac{\pi/2 - \arctan\left(\sqrt{\frac{r}{N-r}}\right)}{\arccos\left(\frac{4km}{N^2} - \frac{4r}{N} + \Gamma\right)} \right\rceil \tag{14.4}$$

where

$$\Gamma = \sqrt{\frac{8rN^3 + 8kmN^2 - 16rkN^2 + 32rkmN - 16k^2m^2}{N^4}}$$

Equation (14.3) implies that the number of iterations can be approximated by

$$T \approx O\left(\frac{\pi/2 - \sqrt{\frac{r}{N-r}}}{\sqrt{\frac{r}{N}}}\right) = O\left(\sqrt{\frac{N}{r}}\right) \tag{14.5}$$

Further analysis shows (Salman and Baram, 2012, Theorem 2) that the maximal probability of measuring a marked state in $K \cap M$ is approximately 1 when $|K|$, $|M| \ll N$ and N is large.

14.3 Quantum associative memory

Associative memory has been perceived as a mechanism which retrieves a stored information object when probed by a like object. The process of such association may involve pattern completion or pattern correction, which, in the quantum context addressed here, will involve the quantum set intersection mechanism specified earlier.

14.3.1   *Pattern completion*

Let I_M be a phase oracle on a set M, called the memory set, of m n-qubit patterns and let x' be a version of a memory pattern $x \in M$ with d missing bits. It is required to output the pattern x based on I_M and x'. The partial pattern is given as a string of binary values 0 and 1 and some unknown bits marked "?". Denoting the set of possible completions of the partial pattern K and its size k, the completion problem can be reduced to the problem of retrieving a member x of the intersection between two sets K and M, $x \in K \cap M$. For example, let $M = \{0101010, 0110100, 1001001, 1111000, 1101100, 1010101, 0000111, 0010010\}$ be a 7-bit memory set of size 8 and let "0110?0?" be a partial pattern with two missing bits, so the completion set is $K = \{0110000, 0110001, 0110100, 0110101\}$. Pattern completion is the computation of the intersection between K and M, which is the memory pattern 0110100. Pattern completion can use, then, the following procedure:

Iterative pattern completion procedure

Given a memory phase oracle I_M and a pattern $x \in \{0,1\}^n$, which is a partial version of some memory pattern with up to d missing bits.

Create the completion phase oracle I_K.

Apply the intersection set search procedure with I_M and I_K.

14.3.2   *Pattern correction*

Let I_M be a phase oracle on a memory set M of m n-qubit patterns and let x' be a version of a memory pattern $x \in M$ with d faulty bits. We are required to output the pattern x based on I_M and x'. The set K of possible corrections of the faulty pattern consists of all patterns within Hamming distance d from x'. The correction problem can be reduced to the problem of retrieving a member x of the intersection between two sets K and M, $x \in K \cap M$. For example, let $M = \{0101010, 0110100, 1001001, 1111000, 1101100, 1010101, 0000111, 0010010\}$ be a 7-bit memory set and let "0110001" be the input pattern with two possible errors. The correction set K consists of all patterns that are in Hamming distance up to 2 from x'.

Pattern correction should then retrieve the memory pattern 0110100. Employing the quantum intersection mechanism we obtain:

Iterative pattern correction procedure

Given a memory phase oracle I_M and a pattern $x \in \{0,1\}^n$, which is a faulty version of some memory pattern with up to d faulty bits.

Create the correction phase oracle I_K.

Apply the iterative intersection search procedure with I_M and I_K.

A generalization of both the pattern completion and the pattern correction procedures for the case of unknown number of possible corrections is straightforward using quantum search for an unknown number of marked states (Boyer *et al.*, 1996).

14.4   Associative memory capacity results

The derivation of most of the following results follows directly from earlier quantum computational analysis (Salman and Baram, 2012).

14.4.1   *Time complexity*

As noted above (Eq. (14.5)), the time complexity of the retrieval procedure with either pattern completion or correction abilities is determined by the complexity of the quantum intersection algorithm and the complexity of the completion and correction operators, yielding, in both cases

$$T = O\left(\sqrt{\frac{N}{|K \cap M|}}\right) \tag{14.6}$$

iterations.

14.4.2   *Equilibrium storage capacity*

The equilibrium capacity of the quantum associative memory is

$$M_{eq} = N \tag{14.7}$$

as every memory pattern of size n in a memory set M of size $m \leq N$ is an equilibrium state, that is, if $Q_x = -HI_0HI_x$, $T = \lfloor \frac{\pi}{4}\sqrt{N} \rfloor$ and

$|\Psi(T)\rangle = Q_x^T H |0\rangle$, then

$$\forall x \in M : |\langle x || T \rangle|^2 \to 1 \text{ as } n \to \infty \qquad (14.8)$$

This, as noted before (Salman and Baram, 2012), is a direct consequence of Grover's algorithm (Grover, 1996) and the results obtained by Boyer *et al.* (1996) concerning the ability to find any member of the size N database with probability close to 1.

14.4.3 *Completion capacity*

Given a pattern x', which is a partial version of some memory pattern xc with d missing bits, we seek the maximal memory size, for which the pattern can be completed with high probability from a random uniformly distributed memory set (McEliece *et al.*, 1987; Baram, 1991; Baram and Sal'ee, 1992). The completion capacity is bounded from above by two different bounds. The first is a result of Grover's quantum search algorithm limitations and the second is a result of the probability of correct completion.

Grover's operator flips the marked states around the zero amplitude (negating their amplitudes) then flips all amplitudes around the average of all amplitudes (Biham *et al.*, 1999). Amplification of the desired amplitudes occurs only when the average of all amplitudes is closer to the amplitudes of the unmarked states than to the marked states. This imposes the following upper bound on the memory size:

$$M_{com} < N/2 \qquad (14.9)$$

The following result (Salman and Baram, 2012, Theorem 3) gives an upper bound on the capacity for pattern completion with high probability: An n-bit associative memory with m random patterns can complete up to d missing bits on average when $m \leq 2^{n-d}$ with probability higher than $(v/(e^v - 1)) \sum_{i=1}^{m} v^{i-1}/i^i$, where $v = m/2^{n-d}$.

Figure 14.1 shows that the lower bound on the probability of successful completion with approximation by only three terms of the sum in $(v/(e^v - 1)) \sum_{i=1}^{m} v^{i-1}/i^i$ as a function of v is higher than 75% for all possible sizes of a non-empty memory within the capacity limits. The resulting completion capacity bound $M_{com}(d) = 2^{n-d}$ agrees with the result for the equilibrium capacity, since $M_{com}(0) = 2^n = N = M_{eq}$.

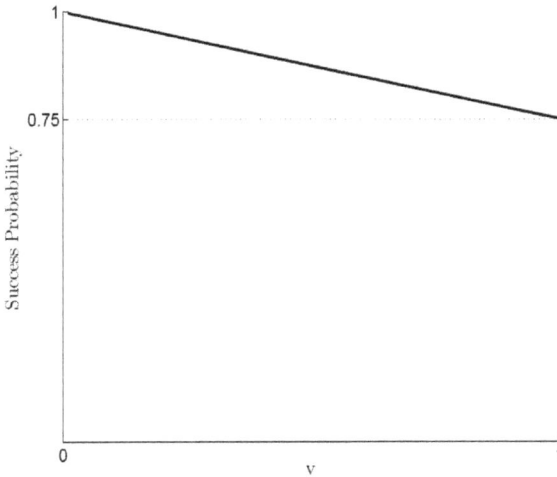

Figure 14.1. Success probability of pattern completion vs. v $\equiv v = m/2^{n-d}$ (Salman and Baram, 2012).

14.4.4 *Correction capacity*

The following result (Salman and Baram, 2012, Theorem 4) gives an upper bound on the capacity for pattern correction with high probability: An n-bit associative memory with m random patterns can correct up to d faulty bits on average when $m \leq 2^{n-d} \binom{n}{d}$ with probability higher than $(v/(e^v - 1)) \sum_{i=1}^{m} v^{i-1}/i^i$, where $v = m/2^{n-d}$.

14.4.5 *Increasing memory size beyond the capacity bound*

The various capacities presented are exponential in n under the assumption $d \ll n$. However, an increase of m beyond the capacity bound results in a decay of the correct completion or correction probability, as depicted in Fig. 14.2. It can be seen that it is more likely to find the correct completion or correction than not to find it as long as $v < 2$. In addition, the model can also output a superposition of a number of possible outputs, by skipping the final measurement operation associated with quantum computation. This is not true for most classical memory models where spurious

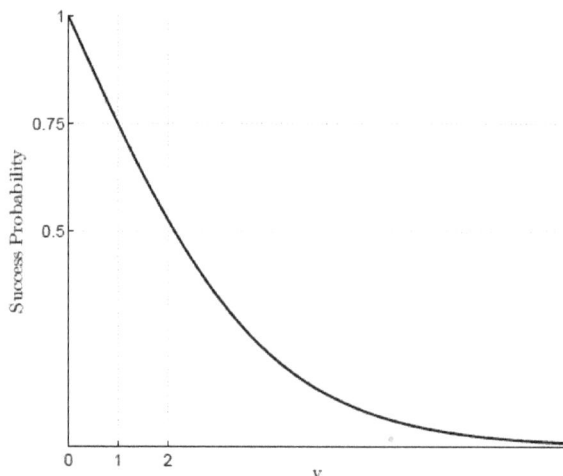

Figure 14.2. Pattern completion or correction probability versus v $\equiv v = m/2^{n-d}$. For $0 < v < 1$, the probability is above 75% and for $v < 2$ is above 50% (Salman and Baram, 2012).

memories arise and the output is usually not a memorized pattern, but, rather, some spurious combination of multiple memory patterns (Hopfield, 1982; Bruck, 1990; Goles and Martínez, 1990).

It is interesting to view the above capacity results from a linguistic graph-theoretic viewpoint. In both the cases of pattern completion and pattern correction, a small value of $v = m/2^{n-d}$, where m is the number of edges, n is the number of nodes and d is the number of incomplete, or incorrect, edges, yields a high probability of successful completion or correction. This is visually illustrated in Figs. 14.1 and 14.2, respectively. As noted in earlier chapters of this book (in particular, Chapter 4), classical considerations suggest that a large n implies a large m (many nodes imply a highly connected graph). Applied to the present quantum computational analysis of neural circuits, the quantum computation results yield a conclusion which, in the contexts of pattern completion and correction, is similar to that reached by classical graph theoretic considerations: small segregated neural circuits, representing short words, are more efficient than large circuits, yielding high capacities of linguistically plausible information.

14.5 Numerical examples and simulations

Following previously presented examples (Salman and Baram, 2012), let us first consider an associative memory of 10 qubits. We have randomly chosen a set of 50 patterns M out of the possible 1,024 to be stored in memory. We also chosen two partial patterns, each with 4 missing qubits, yielding two completion sets K_1 and K_2 of 16 possible completions each. We chose K_1 and K_2 such that they have one and two completions in memory respectively. Figure 14.3(a) shows the memory set, where each vertical line represents a memory pattern, and Fig. 14.3(b) shows the completion set K_1 in the same manner. The amplitudes of the final state of the completion algorithm are shown in Fig. 14.3(c), where the only possible memory completion has amplitude close to 1. Figure 14.4 shows the high amplitudes of the two possible memory completions when the completion set is K_2.

As can be seen, applying our algorithm to both completion sets amplified the states that are possible completions in memory. The amplitudes of the desired states reached up to 96.76% in

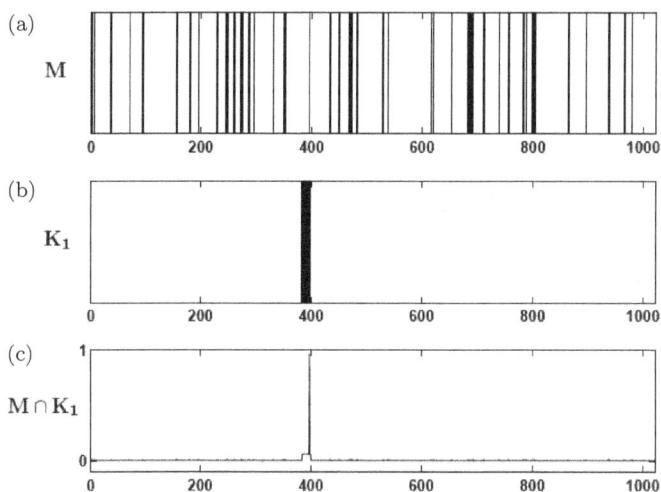

Figure 14.3. (a) A set of memory patterns M (b) a set of possible completions K_1 to a partial pattern, and (c) the memory completion result in amplitudes (Salman and Baram, 2012).

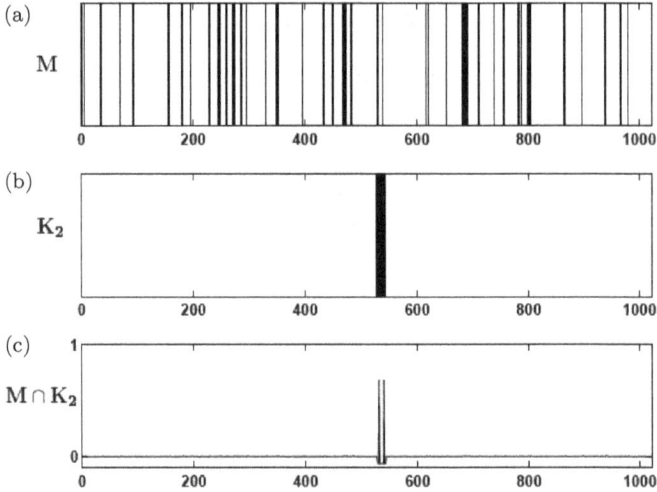

Figure 14.4. (a) A set of memory patterns M (b) a set of possible completions K_2 to a partial pattern, and (c) the memory completion result in amplitudes (Salman and Baram, 2012).

the first case and 68.44% in the second case for each one of the two high amplitudes. Therefore, the probability of measuring the correct completion in the first case is 93.62% and the probability of measuring one of the two correct completions in the second case is 93.67%.

Another simulation was carried out on a 10 qubits associative memory with 27 memory patterns and completion queries with 3 missing bits. The behavior of the different subgroups of the basis states is schematically described in Fig. 14.5 for a series of iterations with the completion operator Q of Eq. (14.3). Each amplitude value indicated represents the amplitudes of all the basis states that belong to the corresponding subgroup. It shows the amplification of states in the intersection group $K \cap M$ and in NOT($K \cup M$) alternatingly, while the amplitudes of states in K\M and M\K stay close to zero.

We have also tested our algorithms with a larger number of qubits in order to verify that the success rate of retrieval grows asymptotically to 1 as the number of qubits grows. For instance, we tested a 30 qubit system with 225 memory patterns and a completion query

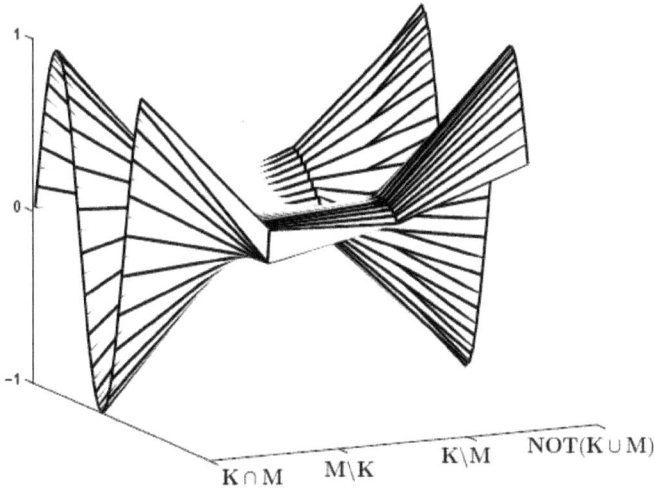

Figure 14.5. Simulation of a series of iterations of the completion algorithm. The graph shows the different behavior of the different subgroups of basis states. K is the completion set and M is the memory set. The memory completions and the non-completions of memories are amplified alternatingly, while the amplitudes of K\M and M\K subgroups stay close to zero (Salman and Baram, 2012).

that has 8 missing bits. We tested different completions of 8 missing bits so that the intersection set size varied from 1 to 10 patterns. Our algorithm found a member of the memory completion set with probability 96.8%. Increasing the memory size to 226 and 227, and thereby bringing the capacity close to its limit resulted in completion probabilities of 93.5% and 86.7%, respectively. Figure 14.6 depicts the success rates of pattern completion in a 30 qubit system. The solid line in Fig. 14.6 depicts the success probability versus the logarithm of the size of memory with completion queries set to 8 missing bits and the number of possible memory completions set to 1. The dashed line depicts the success probability versus the logarithm of the completion query size when the memory size is set to 225 patterns and the number of possible memory completions set to 1. The dotted line depicts the success probability versus the logarithm of the number of possible memory completions when both the memory size and the completion query size are set to 225. The dash-dotted line depicts the success probability versus the number of

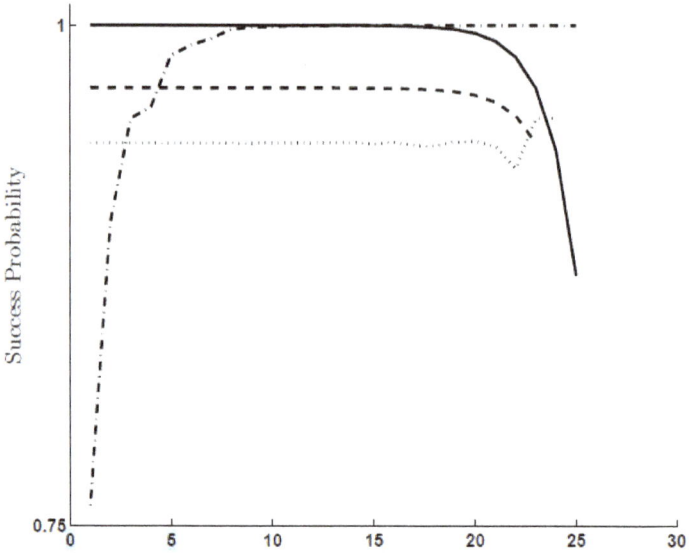

Figure 14.6. Success probability of measuring a desired memory completion versus the log of the memory size (solid), completion query size (dashed), possible memory completions (dotted), and number of qubits (dash-dotted), (Salman and Baram, 2012).

qubits in the system (growing from 5 to 30 qubits) when the memory size, the completion query size and the number of possible memory completions are small constants.

Figure 14.6 shows that the deterioration of the success probability versus the memory size or the completion query size is very slow. For instance, deterioration starts at memory size 2^{26}. Furthermore, the success probability increases when the number of possible memory completions (the size of the intersection set) grows towards the sizes of the completion query and the memory, which indicates that choosing a member of the intersection becomes easy (by randomly choosing a possible completion). Finally, the figure also shows that, as the number of qubits in the system grows, the success probability becomes asymptotically 1, which indicates that, practically, our algorithm produces the intersection when $n \gg 1$.

14.6 Comparison to other works

Quantum computation was previously applied to associative memory by Ventura and Martinez (2000), Ezhov *et al.* (2000), Howell *et al.* (2000), and, subsequently, by others. An algorithm based on the model developed by Ventura and Martinez (2000) was proposed by Arima *et al.* (2008). It was further developed by Arima *et al.* (2009) and analyzed by Miyajima *et al.* (2010). We analyze the two main algorithms (Ventura and Martinez, 2000; Arima *et al.*, 2009) and show their differences with respect to the more recent one (Salman and Baram, 2012). The earlier algorithms are given in the following.

The algorithm proposed by Ventura and Martinez (2000)

Given phase oracles I_M and I_K

Denote $Q_M = -HI_0HI_M$ and $Q_K = -HI_0HI_K$

Let $|\Psi\rangle = \frac{1}{m}\sum_{i=1}^{m}|i\rangle$

Apply $Q_M Q_K$ to $|\Psi\rangle$

Apply Q_K to $|\Psi\rangle$ $T = \left\lfloor \pi/4\sqrt{N/|K \cap M|} \right\rfloor - 2$ times

Measure $|\Psi\rangle$

The algorithm proposed by Arima *et al.* (2009)

Given phase oracles I_M and I_K

Denote $Q_M = -HI_0HI_M$ and $Q_K = -HI_0HI_K$

Let $|\Psi\rangle = \frac{1}{m}\sum_{i=1}^{m}|i\rangle$

Apply $Q_M Q_K$ to $|\Psi\rangle$ T times (T not specified)

Measure $|\Psi\rangle$

The algorithm proposed by Ventura and Martinez (2000) can find only a single marked state with high probability when the memory size m is close to $(N/4) - 2$, as shown by the solid line in Fig. 14.7. The probability of measuring this state reduces by a half when there are two marked states and only one of them is a memory pattern, as shown by the dashed line in Fig. 14.7, and so on.

The algorithm proposed by Arima *et al.* (2009) gives satisfying results only when the memory size exceeds $N/4$, which is exponential

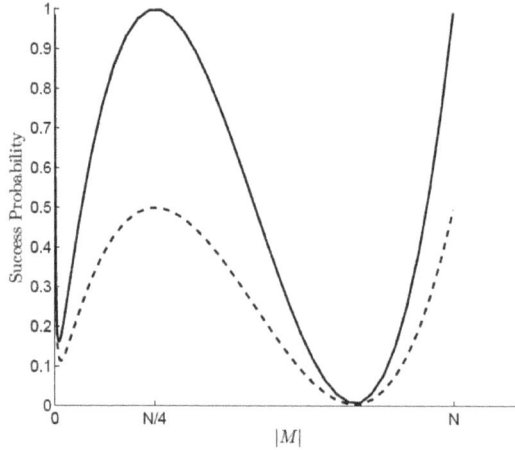

Figure 14.7. Success probability versus memory size $|M|$ for the algorithm proposed by Ventura and Martinez (2000). Optimal results are achieved only when the memory size is close to $N/4$ (Salman and Baram, 2012).

in the number of qubits, leaving the possibility of effective pattern completion only for 2 qubits or less. It is therefore not helpful for associative memory with pattern completion and correction abilities. The success probability of this algorithm versus the memory size is depicted in Fig. 14.8.

Miyajima *et al.* (2010) added a control parameter to tune the algorithm, changing the memory size for which the maximal amplitude is achieved. The algorithm is presented only for one marked state with no completion and correction abilities. The time complexity and stopping criteria were not stated by Arima *et al.* (2008) and were later found to be $O(\sqrt{N})$ (Arima *et al.*, 2009; Miyajima *et al.*, 2010).

In contrast to the two algorithms noted earlier, our pattern completion procedure (Salman and Baram, 2012) presented in Section 14.3.1 achieves high success probability all the way up to memory size $N/4$, as depicted in Fig 14.9. Furthermore, both the earlier two algorithms noted earlier need to initialize the system at a superposition of the memory states

$$|\Psi\rangle = \frac{1}{\sqrt{m}} \sum_{i=1}^{m} |i\rangle \tag{14.10}$$

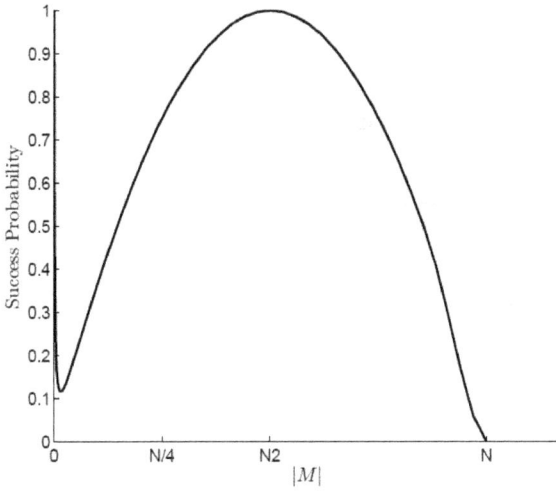

Figure 14.8. Memory size versus success probability in the algorithm proposed by Arima *et al.* (2009). Satisfactory results are achieved only when the memory size exceeds $N/4$ (Salman and Baram, 2012).

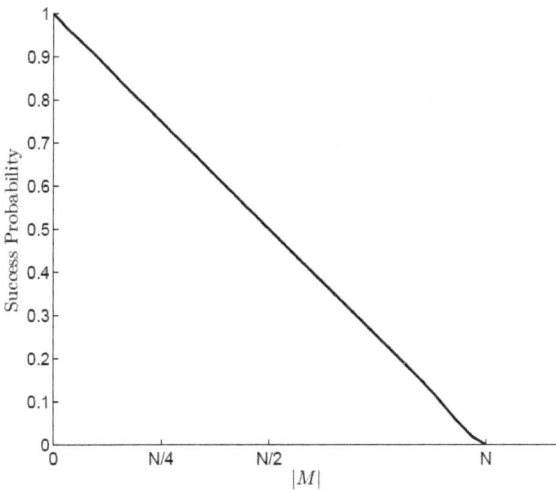

Figure 14.9. Success probability versus memory size $|M|$ in completion and correction procedures (Salman and Baram, 2012).

which presents two difficulties: (i) the time complexity of initialization when the memory size is large and (ii) the need for repeated initialization upon every application of the memory. The latter is important as it adds an exponential factor to the query time, for either completion or correction, and an exponential addition to the single query time when amplitude amplification is needed. Amplitude amplification ensures that we pick the correct pattern with probability 1 by performing the algorithm multiple times. Our algorithm's initialization, on the other hand, does not depend on the memory patterns.

14.7   Discussion

Noting that the main ingredients of quantum search, namely, the Grover iterator (Grover, 1996) and the corresponding phase oracle, seem to be facilitated by neurophysiological processes (Clark, 2014), we have presented here an associative memory model (Salman and Baram, 2012) that might prevail in the cortical domain. The proposed model, applying not only to quantum pattern equilibrium stability but also to quantum pattern completion and quantum pattern correction, is based on a quantum set intersection version of the Grover search algorithm. It has been shown to remove severe limitations on the performance and capacities of earlier propositions based on the Grover principle.

Beyond high information capacities, and considerable reduction in processing time, we have shown, in the context of this book, that the quantum computation approach to associative memory also resolves the linguistic issue associated with large neural networks. Specifically, we have shown that, much like the probabilistic subcriticality associated with random graphs discussed in the first chapters of this book, the quantum approach yields much higher success probabilities for small neural circuits than for large neural circuits. Given the common convention of separation between classical and quantum mathematics, this point of meeting between the two seems to be rather unexpected. As the benefit of quantum computation appears to lie in the transformation of the final measurement into classical mathematics results, we return, in the rest of this book, to the classical domain.

Part IV
Sensorimotor Control

Chapter 15

Circuit Polarity in Sensorimotor Control

15.1 Introduction

The functional role of cortical polarity will be demonstrated in this chapter by examining the correspondence between vision and movement, which will also be the theme of the following two chapters. Vision and movement combine to generate a valuable source of information termed *optical flow*, facilitating accuracy and safety in a fundamental natural behavior, namely, movement. Here, we consider two aspects of such behaviors, obstacle detection and object targeting. While the goals in the two scenarios seem rather contradictory, one aimed at avoiding collision with an object, the other aimed at meeting it head on, the sensorimotor mechanisms governing the two actions will be shown to share key features. Specifically, common to both cases is the process of sensing an expansion of an object's visual image, which is generated by retinal and, accordingly, visual cortex circuit polarization, instigated by local edge detection (Marr and Hildreth, 1980). Moreover, both cases represent a process of circuit firing dynamics, termed *diffusion*, which describes the propagation of sensation-induced neuronal polarization propagating across a neural circuit. Interestingly, the optical flow, or diffusion information, generated in the two cases is used for different calculations. The first being the imminence of collision, which may produce immediate voluntary stoppage; and the second that of the local direction of motion, which will produce an optimal trajectory of motion for visual object targeting.

15.2 Obstacle detection

The detection of an imminent collision is obviously crucial to safe autonomous motion. The underlying sensorimotor control system is generally depicted in Fig. 15.1, where a motion command is cortically transmitted to the motor lobe, which initiates the process of motion. Movement with respect to an imminent obstacle is sensed by the eyes and cortically fed back to the motor system which generates an augmented motion command so as to avoid the obstacle. Range, direction and time to contact can be extracted from the optical flow, generated on the eye retina of a moving observer by the relative motion of the image of a rigid body (Panzeri *et al.*, 2009; Koenderink and van Doorn, 1986; Davies and Green, 1990). Such visual information may be used as a feedback signal to the motor system for generating a collision-free motion trajectory (Baram, 1996). A textured surface induces local divergence in the optical flow, which is inversely proportional to the time to contact (Nelson and Aloimonos, 1989).

The average divergence over the projected image may be obtained by integrating local optical flow measurements along the contour of the projected image, which may be approximated by the state of a locally connected diffusion network (Ringach and Baram, 1992, 1994). Similar networks have been proposed for filling-in of brightness, color and depth in vision (Grossberg and Todorovic, 1988). The resulting average divergence of an object is equivalent to the expansion rate of its projection on the image plane.

The obstacle projection geometry is shown in Fig. 15.2. A focal point O, representing the observer's location, is connected to a point

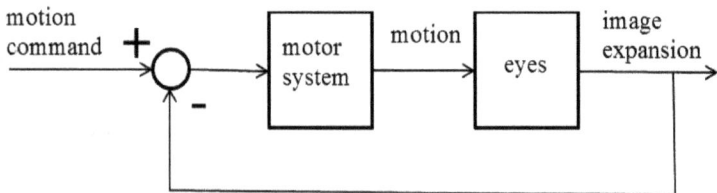

Figure 15.1. Feedback sensorimotor control system.

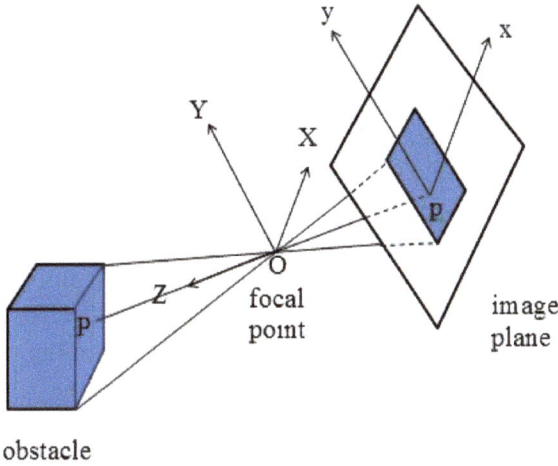

Figure 15.2. Obstacle projection geometry (Baram, 1999).

P on an obstacle. An image plane, representing the retina, is placed at a certain distance from the focal point, perpendicularly to the line OP, which intersects the image plane at point p. The points of the obstacle's boundary are connected to the focal point by straight lines and the projected image of the obstacle is defined by the intersection of these lines with the image plane. Let (X, Y, Z) be a Cartesian coordinate system with origin at the focal point O and let (x, y) be a coordinate system in the image plane, with origin at point p. Let the relative velocity of the observer, located at O, with respect to the obstacle be given by $V = (V_X, V_Y, V_Z)$. Normalizing the velocities, $(\tilde{V}_X, \tilde{V}_Y, \tilde{V}_Z) = (V_X, V_Y, V_Z)/D$, where D is the distance between O and P, the optical divergence at a point p on the contour of the obstacle's projection on the image plane is given by Ringach and Baram (1992, 1994)

$$\Psi(p) \equiv \Delta \cdot v(p) = u_x + v_y = 2\tilde{V}_z + (\tilde{V}_x, \tilde{V}_y) \cdot (F_x, F_y) \qquad (15.1)$$

where $v(p) = (u, v)$ is the velocity of p, and u_x and v_y are the respective partial derivatives in the x and y directions and where F_x and F_y are the slopes of the obstacle face with respect to the X and Y directions. It can be seen that in a steady walk forward at a constant velocity V_z with $V_x = V_y = 0$, the local divergence is inversely proportional to the time to contact with the obstacle $(t = D/V_z = 1/\Psi(p))$. When the observer looks at a point on the obstacle well ahead and walks steadily, both V_z and ∇F are non-zero. Imminent collision will cause abrupt changes in V_z and F_y, increasing both.

As has been observed, the local divergence represented by $\Psi(p)$ is a rather noisy signal, when derived from real images and must be averaged over the projected image of an object (Koenderink and van Doorn,1986; Prazdny, 1983; Nelson and Aloimonos, 1989). For an untextured object, $\Psi(p)$ may be integrated along the contour of the object's image, an operation which can be cortically replaced by a diffusion process (Ringach and Baram, 1992, 1994). Let R denote the projection of the obstacle on the image plane, let $A(R)$ denote the area of the projected image and let ∂R denote the projected boundary of the object. Further let ds and $d\ell$ denote infinitesimal elements of R and ∂R, respectively. Then the average divergence of the points of R is

$$\Phi(R) = \frac{1}{A(R)} \int_{R \setminus \partial R} \Psi(p)ds = \frac{1}{A(R)} \int_{\partial R} v_n(p)d\ell = \frac{1}{A(R)} \frac{dA(R)}{dt}$$

$$(15.2)$$

where $v_n(p)$ is the velocity normal to the edge at p. We see that $\Phi(R)$ can be calculated from the component of the velocity normal to the boundary and that it represents the relative rate of expansion of the obstacle's image. The averaging action in Eq. (15.2) suggests that the noise in $\Psi(p)$ will be largely eliminated and that $\Phi(R)$ will be a more accurate measure of the reciprocal time to contact with the obstacle. As before, it should be noted that, in a steady walk, $\Phi(R)$ represents a certain rate of net optical flow through the obstacle's image, used as a reference signal in the feedback control system. Consequently, the time to contact with the obstacle, as measured by the inverse of the average divergence, reduces approximately at a linear rate, equal to the forward velocity.

This implies that the average divergence, or the net optical flow signal, increases at a rate which is inversely proportional to the forward velocity.

It has been shown (Ringach and Baram, 1992, 1994) that the calculation of $\Phi(R)$ is approximated by the final state of a network of locally connected array of cells, implementing a discrete diffusion process. Similar processes were previously proposed for performing other vision functions (Grossberg and Todorovic, 1988). Assuming a two-dimensional grid arrangement of cells as the one depicted in Fig. 15.3, the cell in position i at time $(k + 1)\Delta t$, where Δt is some time interval, performs the calculation

$$u(i, k + 1) = \frac{1}{|C(i)|} \sum_{p \in C(i)} u(p, k) \quad k = 0, 1, \dots \qquad (15.3)$$

where $C(i)$ is the group of neighboring cells connected to the i'th cell and $|C(i)|$ is the number of cells in that group. This discrete

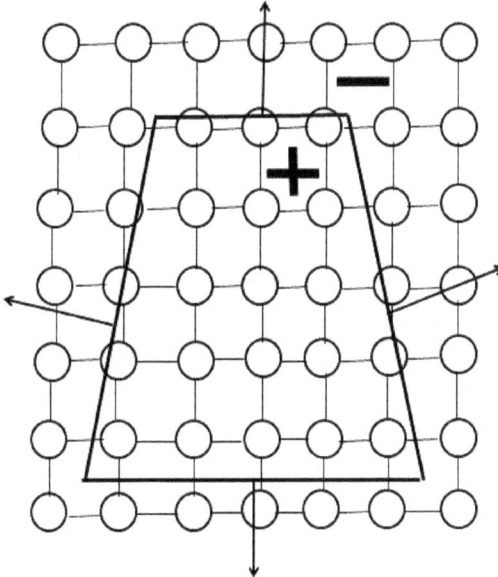

Figure 15.3. A locally connected network for calculating size change by diffusion. Cells within the image boundary have a positive polarity while cells outside the boundary have a negative polarity (Baram, 1999).

diffusion process is initialized by zero everywhere but at the cells on the boundary of the image, which take the values (Marr and Hildreth, 1980)

$$v_n(i) = \frac{I_t(i)}{\sqrt{I_x^2(i) + I_y^2(i)}} \tag{15.4}$$

where I_t, I_x and I_y are the partial derivatives of the image light intensity with respect to t, x and y, respectively.

Cells within the boundary of the image (represented either by retinal receptors or neurons within the visual cortex domain), turned on by the passing edge and forming a circuit of positive membrane polarity, are active, while cells outside the image are silent, having a negative membrane polarity. It has been shown that, for a large number of nodes, the network is resistant even to a high rate (nearly 50%) of connectivity failures (Ringach and Baram, 1992).

Clearly, obstacle avoidance would be achieved by a change of walking pace and/or a change in the direction of movement. Achieving the opposite goal of target encountering is addressed next.

15.3   A bird's eye view on the descent trajectory

When a bird descends towards a point on the ground for the purpose of landing or preying, it should be able to distinguish a three-dimensional object, constituting an obstacle or a target, from its two-dimensional background (Baram, 1996). Extraction of time to contact from optical expansion have been reported and mechanisms for such calculations have been proposed (Nelson and Aloimonos, 1989; Davies and Green, 1990; Ringach and Baram, 1994). The simultaneous expansion of a background image can make target detection difficult. On the other hand, a fast contraction of the background would make it difficult to keep a check of the target location. As we show next, local optimization of the descent scenario facilitates a motion contrast between the optical images of a three-dimensional object and its two-dimensional background, keeping the background contraction rate to a minimum while maximizing the target image expansion, making the distinction between the two relatively easy.

Consider first a three-dimensional target at ground level and an observer, descending along a certain line, as shown in Fig. 15.4. The observer's image of the target is represented by the viewing angle μ, which increases monotonically, as long as the observer's altitude is greater than the height of the target. When the target is replaced by a flat background patch, as shown in Fig. 15.5, the viewing angle of the background patch, θ, will exhibit a different mode of behavior. Denoting the width of the background patch a, its horizontal distance from the observer x, and the observer's altitude h, θ may be written

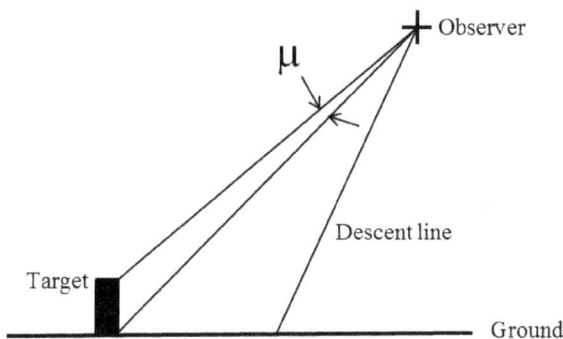

Figure 15.4. Target and observer geometry (Baram, 1996).

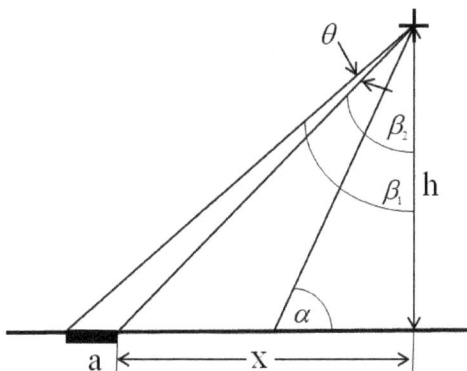

Figure 15.5. Flat background patch and observer geometry (Baram, 1996).

as the difference

$$\theta = \beta_1 - \beta_2 \qquad (15.5)$$

where β_1 and β_2 are the angles between the vertical to the ground at the observer's location and the straight lines connecting the observer to the far and near ends of the patch, respectively. We have

$$\tan \beta_1 = \frac{x+a}{h} \quad \text{and} \quad \tan \beta_2 = \frac{x}{h} \qquad (15.6)$$

hence

$$\tan \theta = \frac{ah}{h^2 + x^2 + ax} \qquad (15.7)$$

Denoting further, $b = x - h/\tan \alpha$, where α is the descent angle, we obtain

$$\frac{1}{\tan \theta} = \frac{1}{a}\left[\left(1 + \frac{1}{\tan^2 \alpha}\right)h + (ab + b^2)\frac{1}{h} + \frac{a+2b}{\tan \alpha}\right] \qquad (15.8)$$

which relates θ to h along the line of descent, yielding a single maximum for θ at

$$h^* = \sqrt{\frac{b(a+b)}{1 + \tan^{-1}\alpha}} \qquad (15.9)$$

The image of the background patch can be made to expand, contract, or remain stationary, by choice of the descent angle α. Contraction or stationarity of the background's image will create a desirable contrast for the target's image, which keeps expanding in the vertical dimension. The greater α is, the faster is the background contraction rate. However, in order to minimize the length of the descent trajectory, α should be as small as possible.

Selecting α locally, so that the size of h along the trajectory, as represented by Eq. (15.9) provides the maximum viewing angle θ of the background patch, will guarantee a maximal view of the background without losing the contrast between the target and its background. It might be further noted that, as in the case of obstacle detection discussed earlier, the image of the target is enhanced by an increase in the number of activated retinal and visual cortex cells, polarized positively by diffusion, as implied by image expansion, whereas the image of the background is progressively, albeit at minimal rate, silenced by negative polarization of the corresponding vision cells.

Taking $h = h^*$ with $b = x - h/\tan \alpha$ and solving Eq. 15.8 for α, we obtain

$$\tan \alpha = \frac{h(1 + 2x)}{x^2 + ax - h^2} \qquad (15.10)$$

The descent trajectory may be calculated for a background patch of a certain standard size, say $a = 1$ (equivalently, all lengths may be assumed, without loss of generality, to be given in units of a). It will then take the form

$$\frac{dh}{dx} = \frac{h(1 + 2x)}{x^2 + x - h^2} \qquad (15.11)$$

which together with the initial condition (x_0, h_0) defines the descent trajectory.

Our analysis was based on the assumption that the local descent angle α is greater than $\tan^{-1}(h/x)$. This is clearly a necessary condition if the observer is to meet the target, with α changing monotonically. It follows from Eq. (15.11) that, for the latter condition to hold at the initial point of the trajectory, we must have $h_0 \leq x_0$. If this is not the case, the ratio x/h must be increased to 1 first. A bird may simply drop straight down to the point $h = x$ letting gravitation alone take effect. Then, spreading wings, it may follow the trajectory represented by Eq. (15.11).

Trajectories are shown in Fig. 15.6 for $h_0 = 50$ with $x_0 = 50$ (red), $x_0 = 100$ (blue) and $x_0 = 200$ (green). In the first case the initial descent is implemented by a drop straight down. It can be seen that in all cases a nice descent trajectory, ending at the target location, is attained.

It might be noted that the trajectory that maintains a constant viewing angle of the background patch in a global sense is the relevant part of the circle which passes through the initial location of the observer and the two ends of the patch. There are, however, two major problems associated with flying along such a circular trajectory: it cannot be calculated locally from optical measurements

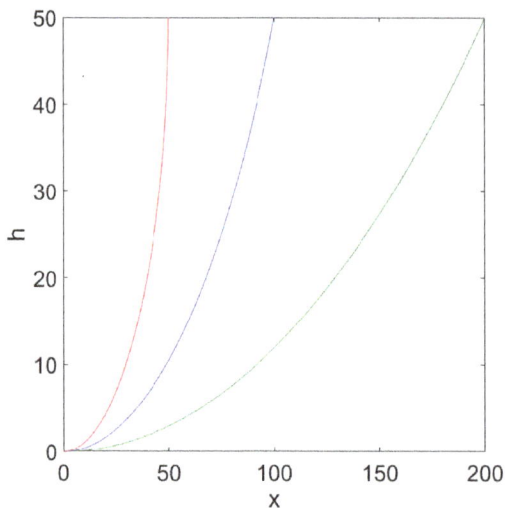

Figure 15.6. Descent trajectories for different initial conditions (Baram, 1996).

alone, and it intersects the ground at a non-zero angle (which might have a disagreeable effect on any bird). In contrast, the proposed trajectory can be derived locally by simply varying the descent angle until the viewing angle appears stationary. In terms of Fig. 15.6, the boundary of the image is replaced by that of the background patch viewed by the bird, and the diffusion process, causing changes in cell polarity stops. Such local stationarity allows a gradual decay of the viewing angle. More importantly, the descent angle safely nullifies at zero altitude.

15.4 Discussion

Movement execution is perhaps the most fundamental task of the brain. Movement has been associated with enhanced neurogenesis, as well as learning and memory (van Praag *et al.*, 1999). Examining two specific natural movement tasks, one aimed at obstacle avoidance, the other aimed at object targeting, we have shown that neural circuit polarity, controlled by the optical flow generated by movement, yields biologically-driven mechanisms for performing both tasks. At the heart of both mechanisms — controlling safe and accurate movement by visual feedback — is the neuron's ability to change its polarity, turning a seemingly complicated computational task into a simple natural reality. It is the realization of the underlying natural process which has made it possible to improve movement in the neurologically impaired, as described in the following sections.

Chapter 16

Autonomous Gait Entrainment in the Neurologically Impaired

16.1 Introduction

The previous chapter presented a cortical process which, by diffusive neuronal polarization, propagates sensory information to achieve a natural movement objective, e.g., obstacle avoidance or prey targeting. We now show that harnessing such abilities can produce significant medical benefits for the neurologically impaired. A seminal study (Martin, 1967) and subsequent clinical tests (Bagley *et al.*, 1991; Morris *et al.*, 1996; Azulay *et al.*, 1999) have shown significant improvement in gait parameters of patients with Parkinson's disease (PD) walking over transverse lines drawn on the ground. Early attempts to generate such visual cues by virtual means have resulted in open-loop systems that impose a constant walking speed on the patient by a constantly moving visual cue (Prothero, 1993; Weghorst *et al.*, 1994; Riess and Weghorst, 1995) or a geometrically patterned treadmill (Hanakawa *et al.*, 1999). A comparison of open-loop visual cuing by virtual means to transverse line markings on the ground (Griffin *et al.*, 2011) has found the first to have a marginal effect and the second to have a significant positive effect on gait parameters in PD patients. Subject-stationary visual cues (Lewis *et al.*, 2000) represent another unnatural imposition where the image moves in the same direction as the subject, which stands in contrast to the opposite motion of an earth-stationary image. External open-loop auditory entrainment of gait — where a rhythmic sound is imposed on the patient in a metronome-like fashion — has been studied

(Kritikos *et al.*, 1995; Thaut *et al.*, 1996; McIntosh *et al.*, 1997; Howe
et al., 2003; Niewboer *et al.*, 2007) and reasoned on neurobiological
grounds (Thaut *et al.*, 2014). Yet, there is a need for constant
vigilance and attention strategies to prevent reversion to impaired
gait patterns caused by repetitive stimuli (Morris *et al.*, 1996). It
has been further suggested that external rhythmic entrainment insti-
gates dopamine reward (Miendlarzewska and Trost, 2014; Salimpoor
et al., 2015) that supports open-loop intervention. However, open-
loop control is known to be inherently unstable, with potentially
disastrous consequences due to error accumulation (Kuo, 1981).
In sharp contrast, closed-loop feedback systems can regulate and
stabilize otherwise unstable dynamics (Kuo, 1981) when correctly
designed. Analysis has shown that walking over earth-stationary
visual markings constitutes a closed-loop feedback control system
which stabilizes and regulates gait (Baram, 1999). Portable virtual
realization of autonomous gait entrainment by sensory feedback,
facilitated by the gradual miniaturization of sensing, computation
and display technologies, has led to the development of a closed-loop
sensory feedback device for gait improvement in the neurologically
impaired (Baram, 2004). While the effect created by the device
is similar to that created by walking on a real floor with visual
markings, the virtual earth-stationary visual markings are location-
independent as the patient is, in principle, free to go anywhere,
making such entrainment completely autonomous.

We employ analytic and experimental findings in demonstrating
and explaining the fundamental differences between external and
autonomous entrainment. From a control-theoretic viewpoint, the
difference between open and closed-loop control systems entails
the difference between dynamic stability and instability. From the
viewpoint of human cognition, the difference between dependence
and independence is the difference between apathy and regression,
on the one hand, and novelty and vigilance, on the other. Reviewing
clinical studies of external and autonomous gait entrainment in
neurological patients, we employ the control theoretic and the human
cognition contexts to explain the differences between the resulting
gait parameters. Significant gait improvement by autonomous visual

and auditory entrainment in patients with PD, typically suffering from basal ganglia dopamine depletion, asserts previous findings that pre-disease learning of reward-seeking behavior can replace actual reward. This effect is also expressed by residual gait improvement, lasting beyond actual entrainment. A change from one-dimensional (transverse lines) to two-dimensional (checkerboard tiles) geometry in autonomous visual entrainment, shown to produce higher novelty and vigilance, is found to result in a particularly high gait improvement in patients with Multiple Sclerosis (MS). We close with a review of clinical studies on autonomous gait entrainment in a variety of neurological disorders, specifically, PD, MS, Cerebral Palsy (CP), Senile Gait (SG) and Previous Stroke (PS). These studies show, on average, pronounced improvement in gait, demonstrating the transformative nature of autonomous gait entrainment across neurological disorders.

16.2   The edge of autonomy in visual feedback

An examination of the natural sensorimotor control system underlying human locomotion with respect to visual scenery reveals that it is the physical motion of the body which generates the visual cue and not the other way around (Baram, 1999). This observation is crucial to understanding the difference between external entrainment and autonomous entrainment of gait and its cognitive effect on movement (Baram, 2009). The first represents a control system operating in open-loop, the second a closed-loop feedback control system. The two control paradigms are illustrated in Fig. 16.1. In external entrainment, a visual cue moving at a constant speed, and/or a rhythmic auditory cue are generated artificially and fed through the eyes and/or the ears to the cortical motor lobe, which activates the limbs so as to respond to the sensory cues. On the other hand, in autonomous entrainment an artificial realization of the natural sensorimotor control system is a closed-loop feedback system, where the generation of the sensory feedback cue is controlled and regulated by the body movement caused by locomotion. The resulting motion of the visual cue and the rhythm of the auditory cue are matched to the motion of the body, which is normally matched to

External entrainment

Autonomous entrainment

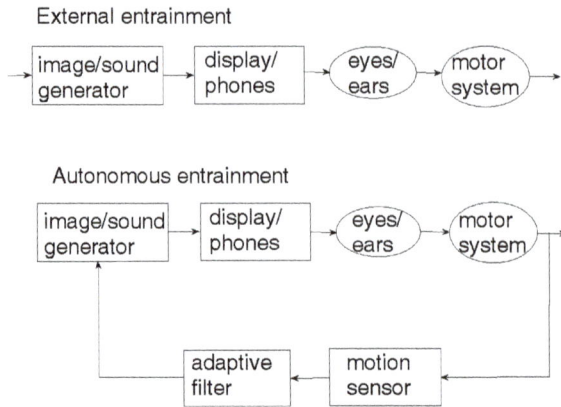

Figure 16.1. Autonomous versus external entrainment of gait: the difference between open and closed-loop sensorimotor control.

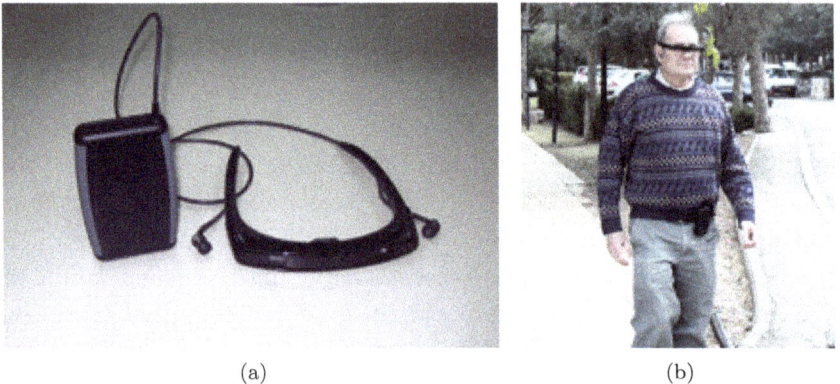

(a)   (b)

Figure 16.2. (a) Autonomous entrainment device and (b) walking with the device (Baram, 2009).

the steps performed by the legs. When there is no motion of the body, there is no sensory cue.

Depicted in Fig. 16.2, a wearable sensory feedback device, employing inertial sensors, an adaptive filter and a microprocessor contained in a belt-mounted cellphone-size box, is connected to a see-through micro-display, generating an earth-stationary visual feedback cue in response to the body motion. In addition to the visual feedback cue delivered by the display, the device also produces an auditory feedback cue in the form of a clicking sound delivered through earphones in response to every step taken by the patient. In contrast to open-loop, metronome-like devices, which attempt to impose a walking pace on the patient by a constant auditory cue, the feedback device produces an auditory cue matched to the walking pattern. A balanced steady walk will generate a rhythmic auditory cue. Any deviation from such a gait pattern will result in a deviation from the auditory rhythm and will be corrected by a change of the gait pattern in a feedback fashion. The head-mounted display and earphones bring the sensory feedback signals closer to the sensors — the eyes and the ears, making the sensory effect more pronounced, easier to follow and to learn. An open-loop capability, producing constant movement of the visual cue and a constant rhythmic auditory cue, was also added to an early version of the device for experimental comparison purposes.

The safety of autonomous gait entrainment is ratified by control theoretic considerations: closed-loop feedback systems are inherently stable when correctly designed due to error correction, while open-loop, externally driven, systems are inherently unstable hence unsafe, due to error accumulation (Kuo, 1981). As can be seen from Fig. 16.1, while error correction (hence, stabilization) is made possible by the sensory feedback path of autonomous entrainment, it cannot be implemented by the open-loop (no feedback) system of external entrainment.

It has been suggested by Jenkinson and Brown (2011) that the level of beta frequency activity in basal ganglia neurons provides a measure of the likelihood that a new voluntary action will be actuated. The frequency level of this activity is modulated by the level of dopamine discharge at sites of cortical input, which are in turn modulated by salient internal and external cues. These insights suggest that autonomous entrainment of gait is more highly rewarded by dopamine discharge than rhythmic external cues. It is interesting, then, to compare clinical findings on the effects of external rhythmic entrainment and autonomous entrainment of gait in patients with PD, suffering from dopamine depletion in basal ganglia neurons.

A study of visual entrainment of gait in patients with PD (Baram *et al.*, 2002) has compared the effects of external (open-loop) constant entrainment and autonomous (closed-loop) feedback entrainment. Patients were off their regular medication (hence, without dopamine enhancement drugs) for 12 hours. The study found that patients who used external entrainment improved their gait on average by 13.8% in walking speed and by 15.0% in stride length (for comparison purposes, we present only first-order statistics). Two of the patients went into freezing of gait midway when using external entrainment. Patients who used autonomous entrainment improved their gait on average by 25.7% in walking speed and by 30.8% in stride length. In addition to doubling the level of improvement in gait parameters with respect to external entrainment, none of the patients employing autonomous entrainment experienced freezing of gait. A subsequent study of autonomous visual gait entrainment in patients with PD on their regular medication (Badarny *et al.*, 2014) found similar on-line and short-term (following 15 min rest and then walking without the device) residual improvement. Residual effects of visual feedback entrainment by markings on the ground have been reported as well (Morris *et al.*, 1996).

It has been suggested that external open-loop entrainment insti-gates dopamine reward (Miendlarzewska and Trost, 2014; Salimpoor *et al.*, 2015), seemingly supporting open-loop intervention. However, as shown by the earlier mentioned clinical results, external entrain-ment also results in adverse effects despite having some positive effects on some of the patients, such as freezing of gait in patients with PD. These mixed results may be characterized as the "double-edged sword" of external entrainment.

16.3 The edge of auditory feedback

Application of external, open-loop, rhythmic (metronome-like) audi-tory entrainment (Nieuwboer *et al.*, 2007) was found to increase or decrease walking speed in patients with PD on their regular medication schedule according to changes in the rhythm frequency above or below their preferred pace, respectively. At the same time, stride length has not been found to be significantly affected by such changes, or by external auditory cuing altogether. At the preferred pace, open-loop rhythmic auditory cuing has been reported to result in 4.2% average improvement in the posture and gait (PG) score, combining walking speed and stride length among a variety of posture and gait measures (Nieuwboer *et al.*, 2007). A study of the effects of autonomous auditory entrainment on gait in patients with PD on their regular medication schedule (Baram *et al.*, 2016) found on average 12.37% improvement in walking speed and 4.30% improvement in stride length with respect to baseline performance. This study has also found a pronounced short-term residual effect of auditory autonomous entrainment in both walking speed (9.1%) and stride length (6.5%). In contrast, external open-loop rhythmic auditory entrainment has been found to have no functional carry-over effects (Nieuwboer *et al.*, 2007).

16.4 Dopamine reward and reward-seeking: The edge of memory

It has been hypothesized that dopamine neurotransmission, highly evidenced in limbic areas of the basal ganglia (Di Chiara *et al.*, 1992), elicits exhilaration or excitement in response to novel stimuli (Cloninger, 1987). The dopamine receptor gene D4DR has been linked to novelty seeking behavior (Ebstein *et al.*, 1996; Benjamin *et al.*, 1996). In the context of cognition, novelty has been associated with "the ability to think and act independently" (The Free Dictionary, 2018). While it has been suggested that seeking novelty — representing intellectual curiosity, aesthetic sensitivity and risk taking — is low in dopamine-deficient patients with PD (Menza *el al.*, 1993), highly pronounced novelty-seeking behavior has been found in patients with PD, regardless of medication status (Djamshidian *et al.*, 2011). Dopamine reward has also been associated with vigilance (Aston-Jones *et al.*, 1994; Ikemoto, 2007), defined as "alert watchfulness" (The Free Dictionary, 2018). A highly insightful study (Jenkinson and Brown, 2011) suggests that the frequency level of beta-range neuronal activity in the basal ganglia is modulated by the level of dopamine discharge at sites of cortical input, which are, in turn, modulated by salient internal and external cues. Moreover, it suggests that the level of the beta-range frequency provides a measure of the likelihood that a new voluntary action will be actuated. Put in our context, this insight suggests that autonomous entrainment of gait is more highly rewarded by dopamine discharge than rhythmic external cues. Yet, dopamine depletion in basal ganglia neurons of patients with PD suggests that it may be reward-seeking, rather than actual dopamine reward, which comes into play in gait entrainment of patients with PD.

While movement has been widely associated with reward, specifically, dopamine discharge by basal ganglia neurons, movement improvement in PD patients suggests the existence of a bypass mechanism. Prior reward learning, possibly facilitated by earlier normal pre-conditioning or medication, provides response incentive without actual dopamine signaling (Wassum *et al.*, 2011). In other words, learning to select responses that lead to reward results in reward-seeking behavior (Graef *et al.*, 2010; Moustafa, 2010). As novelty and vigilance have been widely recognized as key proponents of dopamine reward, the level of dopamine reward or reward-seeking should correspond to the levels of novelty and vigilance. While neither novelty, nor vigilance, have generally accepted measurable manifestations, both seem to be intuitively well understood notions. In a cognitive perspective, novelty has been characterized as the ability to think and act independently (The Free Dictionary, 2018). In this respect, novelty does not seem to apply to external entrainment, which enforces a rhythmic auditory cue, or a constantly moving visual cue, on the patient. Indeed, it has been noted that the failure of external entrainment to produce satisfactory gait control in most patients is caused by the perpetual motion of the image, which, being unaffected by the patient, is eventually neglected (Morris *et al.*, 1996). In sharp contrast, autonomous entrainment, allowing the patient to decide if and when to take a step and at what stride length and speed of walk, is an embodiment of independent, free-willed motion. Yet, vigilance appears to play a role in both external and autonomous entrainment, as both motivate a certain correspondence between step preparation and execution, on the one hand, and the presentation of a sound or an image, on the other. The unequal balance between novelty and vigilance in the two cases suggests that instigating both novelty and vigilance may present a much stronger case than external entrainment which instigates vigilance only, although both external and autonomous entrainments present cases for dopamine reward, or reward-seeking, autonomous entrainment.

16.5 The edge of geometric dimension

Figure 16.3 shows the earth-stationary transverse lines geometry (a) and the checkerboard tiles geometry (b), with shoe markings. These indicate that the patient, by controlling stride-length, has reached a steady-state in which each step ends with a foot placed either between two consecutive transverse lines (in case (a)) or on a tile of a given color (white in case (b)). As the patient, employing either geometry for autonomous visual feedback, has the freedom to move on the floor in any direction, novelty will be visually expressed by edge-crossing or its avoidance. In the case of the transverse lines, edge-crossing can occur in the longitudinal direction only — perpendicularly to the direction of the transverse lines. On the other hand, in the case of the checkerboard tiles, edge-crossing can occur in two orthogonal directions: longitudinal and lateral. Novelty in the first case is one-dimensional, while in the second case it is two-dimensional. The vigilance required in the case of transverse lines is similarly one-dimensional, while the vigilance required in the case of checkerboard tiles is two-dimensional. A dopamine reward, if available, can be expected to be higher in the case of checkerboard tiles than in the case of transverse lines. Simply compare the perceived pleasure of drawing in one dimension to the one of drawing in two dimensions. Following the positive effects found in patients with MS subject to autonomous entrainment (Baram and Miller, 2006, 2007), a clinical study has compared the effects of walking over transverse lines to the effects of walking on checkerboard tiles in such patients (Baram and Miller, 2010). Two groups, each consisting of ten randomly selected patients on their regular medication were tested, one employing transverse lines, the other checkerboard tiles. The study found that while the average improvement with respect to baseline performance in the group employing transverse lines was 7.79% in walking speed and 7.20% in stride length, the average improvement in the group employing checkerboard tiles was 21.09% in walking speed and 12.99% in stride length.

It can be seen that the level of improvement in gait parameters, due to higher-dimensional visual feedback geometry found in patients with MS, increased significantly when the transverse lines geometry was replaced by the checkerboard tiles geometry. This can be attributed to the higher dimensionality of the novelty and the vigilance associated with the checkerboard tiles geometry. These findings were sufficiently conclusive for us to drop any further employment of the transverse lines geometry and adopt the checkerboard tiles geometry in all subsequent trials.

16.6   The edge of polarity

As explained in Section 10.3, the image of a flat visual object placed on the ground can have a significant effect on movement. A tiled floor provides particularly rich information for gait entrainment in the neurologically impaired. Consider the tiled floor image depicted in Fig. 16.3(b). Walking across the tiles generates repeated visual edge crossing, which is detected by retinal sensors. While edge crossing initiates a retinal diffusion process, the color contrast between alternating tile images maintains different polarities of the corresponding retinal circuits, as depicted in Fig. 16.3(b). The visual polarity difference, transmitted from visual to motor cortical region, becomes a motor driving signal which, in turn, results in further gait entrainment. Disturbance in visual rhythm, as evidenced in first step depicted in Fig. 16.3(b), translated into disordered retinal diffusion, are eliminated by correction of gate pattern (represented by subsequent steps in Fig. 16.3(b)).

(a)

(b)

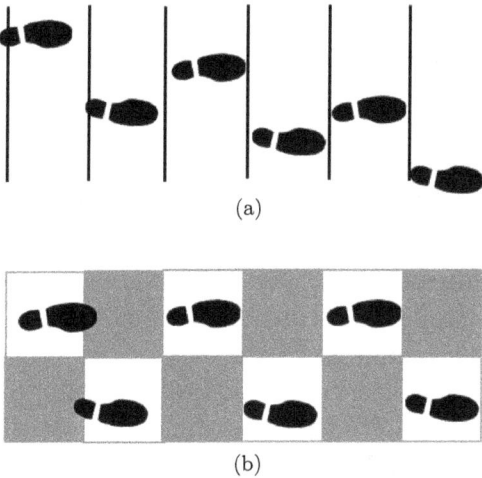

Figure 16.3. Walking with transverse lines (a) and with checkerboard tiles (b) feedback. Following a few initial steps, the steps synchronize with the lines or the tiles. The transition from one-dimensional (a) to two-dimensional (b) novelty and vigilance results in lateral drift elimination.

16.7 Transformative autonomous entrainment of gait across neurological disorders

Further studies on autonomous entrainment in patients with PD have found on-line and residual therapeutic effects following training with combined visual and auditory feedback (Espay *et al.*, 2010; Espay, 2010), effects of "on" pre-dominant freezing of gait (Chong *et al.*, 2011), on-line (Espay *et al.*, 2013; Badarny *et al.*, 2014) and short-term residual (Badarny *et al.*, 2014) improvement by separate visual feedback, freezing predictability by gait initiation with visual feedback (Chong *et al.*, 2015), and gait improvement by combined visual and auditory feedback in patients subject to deep brain stimulation (Souza *et al.*, 2015). Studies of the effects of visual (Baram and Miller, 2006) and auditory (Baram and Miller, 2007) autonomous entrainment on patients with MS have found significant on-line and residual improvement in gait parameters.

Our exploratory studies in PD and MS, suggesting the transformative nature of autonomous entrainment (Baram, 2013b, 2017c), have led to studies of autonomous entrainment in other neurological disorders. Such disorders do not appear to have been approached by technologically-based gait entrainment before, apparently because they present a considerably larger variety of behaviors than patients with PD, and even patients with MS. Consequently, these disorders could not be addressed by external rhythmic entrainment but seemed to present a case for individual self-generated autonomous closed-loop feedback entrainment. The results of our clinical findings for autonomous entrainment in these disorders are now briefly reviewed.

Cerebral palsy (CP) has predominantly pre-natal causes and is symptomatically addressed at a young age. A study of patients with gait disorders due to CP (Baram and Lenger, 2012) found that, for patients training with visual feedback, the short-term residual improvement was 21.7% in walking speed and 8.72% in stride length. For CP patients training with auditory feedback, the short-term residual improvement was 25.43% in the walking speed and 13.58% in the stride length. Age-matched controls who trained with either visual or auditory feedback showed no improvement in gait parameters. A study of randomly selected old-age home residents suffering from senile gait (SG) without PD (Baram *et al.*, 2010) showed that, in patients with baseline performance above the median, the average on-line improvement when using visual feedback was considerably higher (6.31% in walking speed and 6.41% in stride length) than in patients with baseline performance below the median (−0.72% in walking speed and 3.39% in stride length). Improvement in gait parameters was further shown to increase with the number of years of schooling. Patients with SG also suffering from PS (Baram *et al.*, 2010), using on-line visual feedback, improved their walking speed or stride length or both by more than 10% on average. In patients with baseline performance above the median, improvement was considerably higher (13.2% in walking speed and 16.6% in stride length) than in patients with baseline performance below the median (−9.9% in walking speed and −7.7% in stride length).

16.8 Discussion

Put in the context of cortical polarity, the findings reviewed and compared in this chapter unambiguously show that autonomous entrainment of gait by visual and/or auditory feedback is not only decisively safer, but also significantly more effective than external entrainment. Specifically, it is the difference in retinal sensor polarity generated by visual images which entrains the motor system in a feedback fashion. While external gait entrainment is a "double-edged sword" producing adverse effects such as freezing in patients with PD, autonomous entrainment comes without such adverse effects. While the motivation underlying the effectiveness of autonomous entrainment is dopamine reward, it is materialized in patients with PD lacking in dopamine reward by preconditioned or learned reward-seeking behavior. Underlying reward and reward-seeking are the cognitive attributes of novelty and vigilance. Novelty, cognitively characterized as the ability to think and act independently, can only materialize under autonomous entrainment and not under externally enforced entrainment. Geometrically defined with respect to edge crossing under visual feedback, both novelty and vigilance are more highly expressed in movement with respect to two-dimensional visual objects (notably, checkerboard tiles) than in movement with respect to one-dimensional visual objects (notably, one-dimensional transverse lines).

External entrainment represents an open-loop control system which is inherently unstable hence, unsafe for patient use due to error accumulation while autonomous entrainment constitutes a closed-loop feedback control system which is, by error correction, inherently stable and, hence, safe for patient use. Autonomous entrainment of gait by visual or auditory feedback produces significantly higher improvement in gait parameters than external entrainment. The level of improvement in gait parameters due to autonomous entrainment found in PD patients without dopamine supplement was higher than that achieved with dopamine supplement, indicating that the effect of learned reward-seeking behavior can be at least as high as that of actual dopamine reward. While external auditory entrainment of gait in patients with PD produced no residual effects, autonomous auditory entrainment in such patients has resulted in residual improvement, indicating that autonomous entrainment is more effective in learning reward-seeking behavior than external entrainment. The checkerboard tiles geometry, being two-dimensional, produces in autonomous entrainment significantly higher levels of novelty and vigilance, resulting in significantly higher improvement in gait parameters than the one-dimensional transverse lines geometry. The transformative nature of autonomous gait entrainment across neurological disorders, predicted by stability, learnability, reward and reward-seeking considerations, was ratified by gait improvement results obtained in independent clinical studies of patients with different disorders, specifically, PD, MS, CP, SG and PS.

The highly transformative nature of autonomous entrainment implies that patients with different disorders can benefit from it. While PD patients are largely deprived of basal ganglia dopamine production, there does not seem to be any known reason to presume such deprivation in other disorders. Yet, the level of gait improvement can vary significantly, not only with respect to the nature of the disorder, but also with respect to disorder severity and patient's own biological and biographical attributes, such as age, education, mental state and cognitive abilities. While the association with certain attributes, such as education, has been noted, attempting to predict patient's response to autonomous entrainment according to the various attributes would present a painstaking task. Yet, certain categories not addressed by the studies reviewed herewith, such as, e.g., PD accompanied by dementia, seem to be particularly interesting. This, and other relevant categorization, such as cognitive abilities, are suggested for future research. Finally, it might be noted that hundreds of publications related to gait entrainment, while making reference to autonomous (or closed-loop) entrainment, have, in fact, addressed external entrainment without making a distinction between external and autonomous entrainment. We find such presentations to be highly, although not necessarily intentionally, misleading. In order to make the distinction between external and autonomous entrainment unambiguously clear, we have only noted works that have made a clear case for one or the other.

Chapter 17

Circuit Polarity and Singularity Segregation in Cortical Recordings

17.1 Introduction

Hanakawa *et al.* (1999) examined cortical activity in patients with PD assisted by treadmill-induced open-loop visual cue for gait entrainment by employing blood-flow measurements. This early study found that the right lateral pre-motor cortex, which is mainly regulated by cerebellar inputs, was activated to a greater extent in PD patients than in age-matched healthy individuals. On the other hand, the healthy individuals activated mainly the supplementary motor area (SMA), which was under-activated in PD patients.

A later study (Velu *et al.*, 2013), employing electroencephalography (EEG) recordings, has examined the cortical activity in patients with PD assisted by closed-loop visual feedback induced by portable augmented-reality device, as described in the previous chapter. This case study has found a considerably higher increase in the beta frequency range of the connections between the occipital and the motor regions, and between the occipital and the parietal regions of a responding patient with PD, compared to a non-responding patient and to an age-matched healthy individual. Revisiting the results of that study, we show here that they concur with the observations on cortical polarity and singularity segregation reported in this book. This is particularly noteworthy in view of the fact that polarity and singularity segregation have been argued in this book on predominantly mathematical grounds. The cortical recordings serve as direct and vivid evidence of not only the validity of cortical polarity and singularity segregation, but also their specific functional role in the directional connectivity between brain areas associated with the cortical functions of interest, specifically, vision, navigation and movement.

17.2   Recorded cortical sensorimotor activity

Electroencephalography is an electrophysiological monitoring method to record electrical activity of the brain. Cortical activity, relating neural firing-rate to scalp-localized potential values, was recorded employing a portable EEG device and compared in a cortical network composed of occipital (Oz), parietal (P4) and motor (Cz) channels (Velu *et al.*, 2013). The choice of these channels was based on anatomical findings from prior fMRI and PET experiments on paradoxical gait in PD (Hanakawa *et al.*, 1999; Snijers *et al.*, 2011). A patient with PD responding to visual autonomous entrainment (PDr), a non-responding patient with PD (PDnr) and a healthy age-matched individual (Control) participated.

Renormalized partial directed coherence (RPDC), representing directional connectivity between the cortical regions of interest, was used. RPDC provides a scale-free estimator, avoids normalization by outflows and is not restrained by frequency dependent statistical significance (Schelter *et al.*, 2009). The experiment consisted of three walking stages: A (visual feedback turned off), B (visual feedback turned on) and C (visual feedback turned off).

The EEG spectra show color-coded RPDC differences between stages C and A in the delta (0–4 Hz), alpha (8–12 Hz) and beta (12–30 Hz) frequency bands corresponding to the connection between Oz and Cz (Fig. 17.1) and the connection between Oz and P4 (Fig. 17.2). Blue indicates a decrease, while red indicates an increase in the differential RPDC (RPDC(C)-RPDC(A)). It can be seen that there is increased RPDC between Oz and Cz (Fig. 17.1) and between Oz and P4 (Fig. 17.2) in the beta range in PDr, reduced in PDnr and only weakly noticeable in Control. Similar results are seen in the difference in activity between stages B and A, but we only show the differences in RPDC between stages C and A to avoid any confound from neural responses to the presence of visual cues.

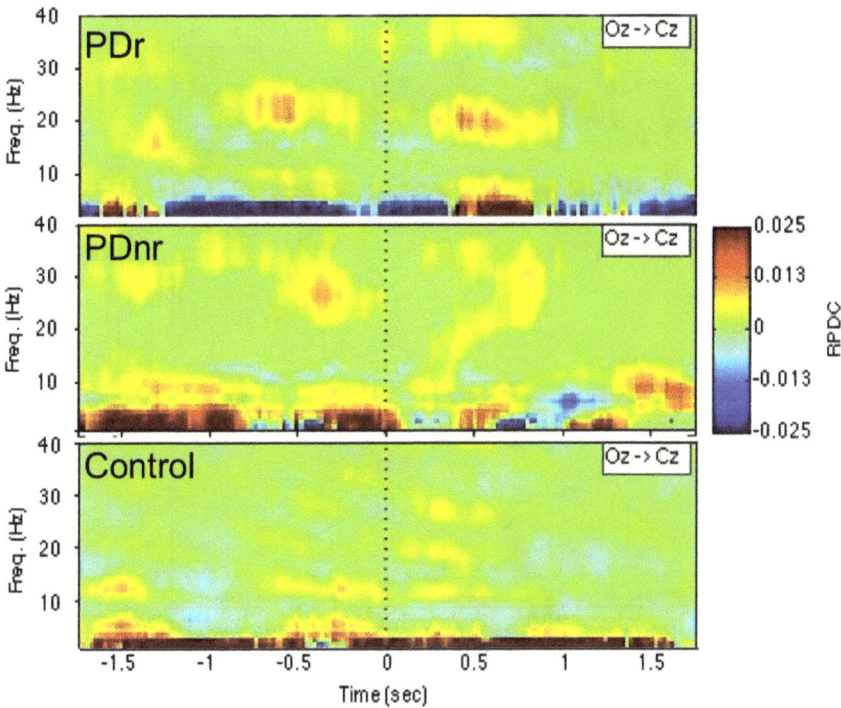

Figure 17.1. Color-coded Renormalized Partial Directed Coherence (RPDC), representing direction-specific connectivity from occipital region (Oz) to motor region (Cz). Blue indicates a decrease in RPDC while red indicates an increase in RPDC between Stage C and Stage A. Increased RPDC appears in the beta frequency range in PDr, reduced in PDnr and only weakly noticable in Control (Velu *et al.*, 2013).

17.3 Circuit polarization and singularity segregation in cortical recordings

Putting the EEG results in the context of cortical polarity, red colored RPDC represents positive polarity in the connectivity between the respective regions (Oz to Cz and Oz to P4), while blue colored RPDC represents negative polarity. It can be seen that in Fig. 17.1, showing Oz to Cz RPDC, there are positive polarity (red-colored) signatures in the beta frequency range (12–30 Hz) corresponding to PDr, considerably reduced in the beta range corresponding to PDnr and almost non-existent in the beta range corresponding

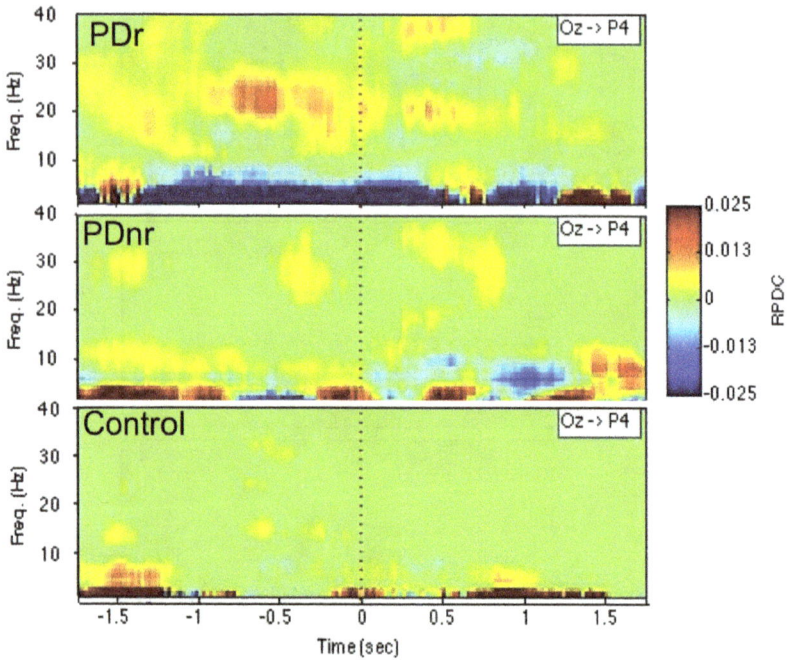

Figure 17.2. Color-coded Renormalized Partial Directed coherence (RPDC), representing direction-specific connectivity from occipital region (Oz) to parietal region (Cz). Blue indicates a decrease in RPDC while red indicates an increase in RPDC between Stage C and Stage A. Occipital region (Oz) to motor region (Cz) flow shows increased RPDC in the beta frequency range in PDr, reduced in PDnr, and only weakly noticeable in Control (Velu *et al.*, 2013).

to Control. On the other hand, there is pre-dominantly negative polarity in the delta frequency range corresponding to PDr, and pre-dominantly positive polarity in the delta range corresponding to PDnr, considerably reduced in the delta range corresponding to Control. There is a very weakly noticeable activity in the alpha frequency range corresponding to all three participants. Essentially similar results can be seen in Fig. 17.2, showing Oz to P4 RPDC. It might be noted that in the beta range for PDr, there are positive (red) RPDC signatures around both 0.5 sec and −0.5 sec in both Fig. 17.1 and Fig. 17.2. These signatures appear to correspond to two consecutive steps, about one second apart within the recorded walking sequence, centered at 0 sec.

In the context of segregated singularity, the visible differences between the RPDC signatures corresponding to the different frequency zones, delta, alpha and beta, representing different connectivity sequences, indicate segregation between the modes of connectivity and, hence, segregation of the corresponding cortical circuits. As shown in this book, a difference in circuit connectivity would imply a difference in circuit firing dynamics, as represented by the EEG activities corresponding to the three frequency domains. In closing, we note again that it was not our purpose here to point out the advantages of visual feedback for improvement of PD gait, which was done in the original study (Velu *et al.*, 2013), but rather to point out the presence of cortical polarization and singularity segregation, mathematically revealed and analyzed in this book, and ratified by the EEG recordings.

17.4   Discussion

Case studies are highly valuable means for ground-breaking research (Solomon, 2006; Carey, 2010; Nissen and Wynn, 2014). Their merits include:

- detecting novelties,
- generating hypotheses,
- applicability when other research design are not implementable,
- allowing emphasis on narrative aspect (in depth understanding) and
- educational values.

The major limitations are:

- lack of ability to generalize, or to establish cause-effect relationship,
- danger of over interpretation,
- retrospective design, and
- distraction by focusing on the unusual.

Results of case studies, while understandably subject-dependent, can be rather conclusive. The results presented and analyzed in this chapter represent a positive case in point. Each of the subjects tested represents a widely recognized category. Among patients with PD, there are patients that respond to treatment and patients that do not. Controls are, normally, age-matched healthy individuals that do not show noticeable response to treatment. As the purpose of the study discussed here was to experimentally examine a theoretically-founded hypothesis regarding, essentially, a model, and, only remotely, a cure for a disease (which has been examined in clinical studies of large cohorts of patients before), a case study was an important first step. The conclusive results were highly ratifying of the theory tested here: cortical activity is driven by circuit polarization and its dynamic singularities are highly segregated. Motivated readers and researchers are highly encouraged to conduct and expand such studies.

Epilogue

Over 100 years of brain research have produced an incredibly large body of knowledge and understanding of this wondrous creation of nature. Experimental studies have addressed many aspects of relevance in considerable detail. Yet, over the years, it has become increasingly clear that putting it all together would require collaboration between a host of scientific disciplines, including biology, chemistry, physics, cognition, psychology and linguistics. While different branches of mathematics appear to be involved in each of these disciplines, it has been suggested that a complete understanding of the mind would require "a new kind of science", yet to be revealed (Penrose, 1994).

Poincare's 19th century discovery of singularities, or attractors, in systems dynamics took nearly a hundred years to materialize as a theory of chaos, dominating all aspects of today's dynamical systems theory. The theories of prime numbers and random graphs represent abstract mathematics in its purest form. While the former, maintaining its somewhat supreme mathematical status, has been slowly finding its isolated ways into certain natural sciences, the latter has been more generous in revealing its benefits in technological domains, such as electronic communication, transportation and computer networks. Both the theories of prime numbers and random graphs have been known to hold, and occasionally reveal, deep and fundamental secrets, repeatedly surprising and astonishing scientifically-minded lay persons, theoreticians and practitioners of the sciences alike. Early 20th century quantum physics, later materializing into a theory of quantum computation, has exponentially expanded the theoretical limits of information capacity. Yet, in terms of scientific research, each of these mathematical disciplines has essentially remained encapsulated within its own stringent borders.

The present book has examined the combined roles of the earlier mentioned mathematical disciplines in the cortical domain. Considered separately, global attractors determine the nature of firing-rate dynamics, consistent with cortical functionality. Subcritical connectivity in randomly generated neural network graphs produces segregated neural circuits. Prime numbers determine separate neural circuit and circuit category sizes, consistent with cortical lobe size limitations. Cortical quantum computation mechanisms enhance the information storage, completion and correction capacities of neural circuits. These mathematical disciplines surface, separate or combined, in a multitude of cortical operations. Circuit size and connectivity define not only firing-rate dynamics, but also memory capacity, circuit category and the maximal lengths of circuit tree firing-rate cycles. Circuit connectivity further determines synchrony or asynchrony of circuit firing. As shown in this book, there is a considerable added value in the cortical synergy of the four mathematical disciplines noted earlier. We have shown that such synergy, representing perhaps, "a new kind of science", can affect cortical operation in a fundamental manner, dominating the activities of the mind.

References

Abbott, L. F. (1994). Decoding neuronal firing and modeling neural networks. *Quart Rev Biophys* 27: 291–331.

Abraham, R. H., Gardini, L. and Mira, C. (1997). *Chaos in Discrete Dynamical Systems*. Springer-Verlag, Berlin.

Abraham, W. C. (2008). Metaplasticity: tuning synapses and networks for plasticity. *Nature Rev Neurosci* 9(5): 65–75.

Abraham, W. C. and Bear, M. F. (1996). Metaplasticity: the plasticity of synaptic plasticity. *Trends Neurosci* 19(4): 126–30.

Adrian, E. D. and Zotterman, Y. (1926). The impulses produced by sensory nerve endings: Part II: The response of a single end organ. *J Physiol* 61: 151–171.

Agrawal, M., Kayal, N. and Saxena, N. (2004). PRIMES is in P. *Annals of Mathematics* 160(2): 781–793.

Aihara, K., Takabe, T. and Toyoda, M. (1990). Chaotic neural networks. *Phys. Lett. A* 144(6,7): 333–338.

Ajazi, F., Chavez–Demoulin, V. and Turova, T. (2019). Networks of random trees as a model of neuronal connectivity. *J Math Biol* 79: 1639–1663.

Amari, S. (1972). Learning patterns and pattern sequences by self-organizing nets of threshold elements. *IEEE Trans Comput.* 21: 1197–1206.

Amit, D. J., Gutfreund, H. and Sompolinsky, H. (1987). Statistical mechanics of neural networks near saturation. *Annals of Physics.* 173(1): 30–67.

Apostol, T. M. (1976). Dirichlet's theorem on primes in arithmetical progressions. In *Introduction to Analytic Number Theory*, pp. 146–156. New York; Heidelberg: Springer-Verlag.

Apostol, T. M. (2000). A centennial history of the prime number theorem. In Bambah, R. P., Dumir, V. C. and Hans-Gill, R. J. (eds), *Number Theory (Trends in Mathematics)*, pp. 1–14. Basel: Birkhäuser.

Arima, K., Miyajima, H., Shigei, N. and Maeda, M. (2008). Some properties of quantum data search algorithms. In *Proceedings of the 23rd International Technical Conference on Circuits/Systems, Computers and Communication (ITC-CSCC2008)*: 1169–1172.

Arima, K., Shigei, N. and Miyajima, H. (2009). A proposal of a quantum search algorithm. In *Fourth International Conference on Computer Sciences and Convergence Information Technology*:1559–1564.

Arnold, V. I. (1989). *Mathematical Methods of Classical Mechanics*. Springer-Verlag.

Ashby, M. C. and Isaac, J. T. R. (2011). Maturation of a recurrent excitatory neocortical circuit by experience-dependent unsilencing of newly formed dendritic spines. *Neuron* 70 (3): 510–521.

Aston-Jones, G., Rajkowski, J., Kubiak, P. and Alexinsky, T. (1994). Locus coeruleus neurons in monkey are selectively activated by attended cues in a vigilance task. *J Neurosci* 14: 4467–4480.

Atwood, H. L. and Wojtowicz, J. M. (1999). Silent synapses in neural plasticity: current evidence. *Learn Mem* 6: 542-571.

Awodey, S. (2010). *Category Theory* (2nd ed.). Oxford University Press.

Azulay, J. P., Mesure, S., Amblard, B., Blin, O., Sangla, I. and Pouget J. (1999). Visual control of locomotion in Parkinson's disease. *Brain* 122(1): 111–20.

Badarny, S., Aharon-Peretz, J., Susel, Z., Habib, G. and Baram, Y. (2014). Virtual reality feedback cues for improvement of gait in patients with Parkinsons Disease. *Tremor & Other Hyperkinet Mov* (NY) 4: 225.

Bagley, S., Kelly, B., Tunniclife, N., Turnbull, G. I. and Walker, J. M. (1991). The effect of visual cues on the gait of independently mobile Parkinson's disease patients. *Physiotherapy* 77: 415–420.

Balice-Gordon, R. J. and Lichtman, J.W. (1994). Long-term synapse loss induced by focal blockade of postsynaptic receptors. *Nature* 372: 519–524.

Balice-Gordon, R. J., Chua, C., Nelson, C. C. and Lichtman, J. W. (1993). Gradual loss of synaptic cartels precedes axon withdrawal at developing neuromuscular junctions. *Neuron* 11: 801–815.

Baram, Y. (1991). On the capacity of ternary Hebbian networks. *IEEE Trans. on Information Theory* 37(3): 528–534.

Baram, Y. (1996). A bird's eye view on the decent trajectory. *IEEE Trans. Aerosp and Electr Sys* 32(3): 1085–1087.

Baram, Y. (1999). Walking on tiles. *Neur Proc Lett* 10: 81–87.

Baram, Y. (2004). Closed-loop augmented reality apparatus. US Patent No. 6,734,834-B1.

Baram, Y. (2009). Closed-loop augmented reality for movement disorders. *Front in Neurosce* 3(1): 112–113.

Baram, Y. (2012). Noninvertibility, chaotic coding and chaotic multiplexity in synaptically modulated neural firing. *Neur Comp* 24(3): 676–699.

Baram, Y. (2013a). Global attractor alphabet of neural firing modes. *J Neurophys* 110: 907–915.

Baram, Y. (2013b). Virtual sensory feedback for gait improvement in neurological patients. *Front Neurol* 4: Article 138.

Baram, Y. (2017a). Developmental metaplasticity in neural circuit codes of firing and structure. *Neur Netw* 85: 182–196.

Baram, Y. (2017b). Asynchronous segregation of cortical circuits and their function: A life-long role for synaptic death. *AIMS Neurosci* 4(2): 87–101.

Baram, Y. (2017c). Transformative autonomous entrainment of gait in neurological patients (Invited Review). *J Neurolog Neurosci* 8(1): 177.

Baram, Y. (2018). Circuit polarity effect of cortical connectivity, activity, and memory. *Neur Comp* 30(11): 3037–3071.

Baram, Y. (2020a). Primal categories of neural polarity codes. *Cogn Neurodyn* 14(1): 125–135. https://doi.org/10.1007/s11571-019-09552-x.

Baram, Y. (2020b). Probabilistically segregated neural circuits and subcritical linguistics. *Cogn Neurodyn* 14(6), 837–848. https://doi.org/10.1007/s11571-020-09602-9.

Baram, Y. (2021). Primal-size neural circuits in meta-periodic interaction. *Cogn Neurodyn* 15(2): 359–367. https://doi.org/10.1007/s11571-020-09613-6.

Baram, Y. and Lenger, R. (2012). Gait improvement in patients with cerebral palsy by visual and auditory feedback. *Neuromodul* 15(1):48–52.

Baram, Y. and Miller, A. (2006). Effects of virtual reality cues on gait in multiple sclerosis patients. *Neurology* 66: 178–181.

Baram, Y. and Miller, A. (2007). Auditory feedback for improvement of gait in multiple sclerosis patients. *J Neurol Sci* 254: 90–94.

Baram, Y. and Miller, A. (2010). Glide-symmetric locomotion reinforcement in patients with multiple sclerosis by visual feedback. *Disab & Rehab: Assis Techn* 5(5): 323–326.

Baram, Y. and Sal'ee, D. (1992). Lower bounds on the capacities of binary and ternary networks storing sparse random vectors. *IEEE Trans on Inform Thy* 38(6): 1633–1647.

Baram, Y., Aharon-Peretz, J. and Lenger, R. (2010). Virtual reality feedback for gait improvement in patients with idiopathic senile gait disorders and in patient with history of strokes. *J Amer Geriat Soc* 58(1): 191–192.

Baram, Y., Aharon-Peretz, J., Simionotici, Y. and Ron, L. (2002). Walking on virtual tiles. *Neur Proc Lett* 16: 227–233.

Baram, Y., Aharon-Peretz, J., Badarny, S., Susel, Z. and Schlesinger, I. (2016). Closed-loop auditory feedback for the improvement of gait in patients with Parkinson's disease. *J Neurol Sci* 363(15): 104–106.

Baroni, F. and Mazzoni, A. (2014). Heterogeneity of heterogeneities in neuronal networks. *Front Comput Neurosci*, https://doi.org/10.3389/fncom.2014.00161.

Bassett, D. S., Meyed-Lindenberg, A., Archard, S., Duke, T. and Bullmore, E. (2006) Adaptive reconfiguration of fractal small-world human brain functional networks. *Proc Nat Acad of Sci* 103(51): 19518–19523.

Battaglia, D., Witt, A., Wolf, F. and Geisel, T. (2012). Dynamic effective connectivity of inter-areal brain circuits. *PLOS Comp. Biol.* 8(3): e1002438.

Bengtsson, I. and Życzkowski, K. (2017). *Geometry of Quantum States: An Introduction to Quantum Entanglement* (2nd ed.), pp. 313–354. Cambridge University Press.

Benjamin, J., Li, L., Patterson, C., Greenberg, B. D., Murphy, D. L. and Hamer, D. H. (1996). Population and familial association between the D4 dopamine receptor gene and measures of novelty seeking. *Nature Genet* 12: 81–84.

Bennett, B. D., Callaway, J. C. and Wilson, C. J. (2000). Intrinsic membrane properties underlying spontaneous tonic firing in neostriatal cholinergic interneurons. *J Neurosci* 20(22): 8493–8503.

Bertels, K., Neuberg, L., Vassiliadis, S. and Pechanek, D. G. (1998). Chaos and neural network learning. Some observations. *Neur. Proc. Lett.* 7: 69–80.

Bick, C. and Rabinovich, M. I. (2009). Dynamical origin of the effective storage capacity in the brain's working memory. *Phys Rev Lett* 103: 218101-1-4.

Bienenstock, E. L., Cooper, L. N. and Munro, P. W. (1982). Theory for the development of neuron selectivity: orientation specificity and binocular interaction in visual cortex. *J Neurosci* 2: 32–48.

Biggs, N., Lloyd, E. and Wilson, R. (1986). *Graph Theory, 1736–1936*. Oxford University Press.

Biham, E., Biham, O., Biron, D., Grassl, M. and Lidar, D.A. (1999). Grover's quantum search algorithm for an arbitrary initial amplitude distribution. *Physical Review A* 60(4): 2742–2745.

Bihan, F. and Dickenstein, A. (2017). Descartes' rule of signs for polynomial systems supported on circuits. *Intern Math Res Notices* 2017(22): 6867–6893.

Blanchard, P., Devaney, R. L. and Hall, G. R. (2006). *Differential Equations*. London: Thompson.

Boklan, K. D. and Conway, J. H., (2017). Expect at most one billionth of a new Fermat prime!. *The Mathematical Intelligencer* 39(1): 3–5.

Bollobás, B. (1984). The evolution of random graphs. *Trans Am Math Soc* 286(1): 257–274.

Bollobás, B. (2001). *The Evolution of Random Graphs — The Giant Component* (2nd ed.), pp. 130–159. Cambridge University Press.

Bollobás, B., Janson, S. and Riordan, O. (2007). The phase transition in inhomogeneous random graphs. *Rand. Struct. Alg.* 31: 3–122.

Bonzon, P. (2017). Towards neuro-inspired symbolic models of cognition: linking neural dynamics to behaviors through asynchronous communications. *Cogn Neurodyn* 11(4): 327–353.

Borg-Graham, L. J., Monier, C. and Frégnac, Y. (1998). Visual input evokes transient and strong shunting inhibition in visual cortical neurons. *Nature* 393(6683): 369–373.

Borisyuk, R., al Azad, A. K., Conte, D., Roberts, A. and Soffe, S. R. (2014). A developmental approach to predicting neuronal connectivity from small biological datasets: A gradient-based neuron growth model. *PLoS One* 9(2).

Borisyuk, R., al Azad, A. K., Conte, D., Roberts, A. and Soffe, S.R. (2011). Modeling the connectome of a simple spinal cord. *Front Neuroinform* 5: 20.

Bourgeois, J. P. (1993). Synaptogenesis in the prefrontal cortex of the Macaque. In B. do Boysson-Bardies (ed), *Develop Neurocog: Speech and Face Processing in the First Year of Life*, pp. 31–39. Norwell, MA: Kluwer.

Bourgeois, J. P. and Rakic, P. (1993). Changing of synaptic density in the primary visual cortex of the rhesus monkey from fetal to adult age. *J Neurosci* 13:2801–2820.

Boyer, M., Brassard, G., Hoyer, P. and Tapp, A. (1996). Tight bounds on quantum searching. In Toffoli, T., Biaford, M. and Lean, J. (eds), *Fourth Workshop on Physics and Computation*, pp. 36–43. New England Complex System Institute.

Brassard, G., Hoyer, P. and Tapp, A. (1998). Quantum counting. In *ICALP '98: Proceedings of the 25th International Colloquium on Automata, Languages and Programming*, pp. 820–831.

Brocket, R. (1970). *Finite Dimensional Linear Systems*. New York: Wiley.

Bruck, J. (1990). On the convergence properties of the hopfield model. *Proc IEEE* 78(10): 1579–1585.

Bruck, J. and Roychowdhury, V. P.(1990). On the number of spurious memories in the Hopfield model. *IEEE Trans Info Theo* IT-36(2): 393–397.

Bruins, E. M. (1974). The recto of the Rhind Mathematical Papyrus. How did the ancient Egyptian scribe prepare it?. *Archive for History of Exact Sciences* 12(4): 291–298.

Brunel, N. and Sergi, S. (1998). Firing frequency of leaky integrate-and-fire neurons with synaptic current dynamics. *J theor Biol* 195: 87–95.

Bullock, T. H. (1997). Signals and signs in the nervous system: The dynamic anatomy of electrical activity is probably information-rich. *Proc Natl Acad Sci U. S. A.* 94: 1–6.

Buzsáki, G. (2010). Neural syntax: cell assemblies, synapsembles, and readers. *Neuron* 68(3): 362–385.

Cajal, R. S. (1890). *Manual de Anatomia Patológica General* (in Spanish).

Calvin, W. H. (1996). *The Cerebral Code: Thinking a Thought in the Mosaics of the Mind*. Cambridge, MA: MIT Press.

Caporale, N. and Yang D. (2008). Spike timing-dependent plasticity: A Hebbian learning rule. *Annu Rev Neurosci* 31: 25–46.

Carandini, M. and Ferster, D. (2000). Membrane Potential and Firing Rate in Cat Primary Visual Cortex. *J Neurosci* 20: 470-484.

Carey, J. C. (2010). The importance of case reports in advancing scientific knowledge of rare diseases. *Adv Exp Med Biol* 686:77–86.

Carlsson, A. and Lindquist M. A. (1963). Effect of chlorpromazine and haloperidol on formation of 3-methoxytyramine and normetanephrine in mouse brain. *Acta Pharmacol Toxicol* 20:140–144.

Cassiday, G. (2013). The Arecibo Message. Available at http://www.physics.utah.edu/~cassiday/p1080/lec08.html (accessed 30 Jan 2021).

Castellani, G. C. (2017). Comparison between Oja's and BCM neural networks models in finding useful projections in high-dimensional spaces. Scuola di Scienze, Dipartimento di Fisica e Astronomia, Corso di Laurea in Fisica, Alma Mater Studiorum, Universita di Bologna.

Castellani, G. C., Quinlan, E. M., Cooper, L. N. and Shouval, H. Z. (2001). A biophysical model of bidirectional synaptic plasticity: dependence on AMPA and NMDA receptors. *Proc Natl Acad Sci U S A.* 98(22): 12772–12777.

Cauchy, A. L. (1813). Recherche sur les polyèdres — premier mémoire. *J. de l'École polytechnique*|9(Cahier 16): 66–86.

Chan, J. (1996). Prime time! *Math Horizons* 3(3): 23–25.

Chase, W. G. and Ericsson, K. A. (1982). Skill and working memory. In Bower, G.H. (ed), *Psychol Learn Motiv* 16: 1–58.

Chechik, G., Meilijson, I. and Ruppin, E. (1998). Synaptic pruning in development: A computational account. *Neur Comp* 10(7): 1759–1777.

Chklovskii, D. B., Mel B. W. and Svoboda K. (2004). Cortical rewiring and information storage. *Nature* 782-8. doi: 10.1038

Chong, R., Lee, K. H., Morgan, J. and Wakade, C. (2015). Duration of step initiation predicts freezing in Parkinson's disease. *Acta Neurologica Scandinavica* 132(2): 105–110. DOI: 10.1111/ane.12361.

Chong, R., Lee, K. H., Morgan, J., Mehta, S., Griffin, J., Marchant, J., Searle, N., Sims, J. and Seth, K. (2011). Closed-loop VR-based interaction to improve walking in Parkinson's disease. *J Nov Physiother* 1: 1–7.

Chou, P. A. (1989). The capacity of the Kanerva associative memory. *IEEE Trans on Info Theory* 35(2): 281–288. https://doi.org/10.1109/18.32123.

Clark, K. B. (2014). Basis for a neuronal version of Grover's quantum algorithm. *Fron Mol Neurosci* 7(29): 1–20.

Cloninger C. R. (1987). A systematic method for clinical description and classification of personality variants. A proposal. *Arch Gen Psychiatry* 44(6): 573–588.

Cohen, M. A. and Grossberg, S. (1983). Absolute stability of global pattern formation and parallel memory storage by competitive neural networks. *IEEE Trans Syst, Man, and Cybern* 13: 815–826.

Colman, A. M. (2008). *A Dictionary of Psychology.* Oxford University Press.

Connor, J. A. and Stevens, C. F. (1971). Voltage clamp studies of a transient outward membrane current in gastropod neural somata. *J Physiol* 213: 21–30.

Constantine-Paton, M., Cline, H. T. and Debski, E. (1990). Patterned activity, synaptic convergence, and the NMDA receptor in developing visual pathways. *Annu Rev Neurosci* 13:129–154.

Cooper, L. N., Intrator, N., Blais, B. S. and Shouval, H. Z. (2004). *Theory of Cortical Plasticity.* New Jersey: World Scientific.

Cormen, T. H., Leiserson, C. E., Rivest, R. L. and Stein, C. (2001). *Introduction to Algorithms* (2nd ed.), pp. 232–236. MIT Press and McGraw-Hill.

Cowan, N. (1995). *Attention and Memory: An Integrated Framework.* Oxford University Press.

Cowan, N. (2001). The magical number 4 in short-term memory: A reconsideration of mental storage capacity. *Behavioral and Brain Sciences* 24(1): 87–114.

Craik, F. and Bialystok, E. (2006). Cognition through the lifespan: mechanisms of change. *Trends Cogn Sci* 10(3): 131–138.

Culican, S. M., Nelson, C. C. and Lichtman, J. W. (1998). Axon withdrawal during synapse elimination at the neuromuscular junction is accompanied by disassembly of the postsynaptic specialization and withdrawal of Schwann cell processes. *J Neurosci* 18(13): 4953–4965.

Cymbalyuk, G.S. and Shilnikov, A.L. (2005). Coexistence of tonic spiking oscillations in a leech neuron model, *J Comp Neurosci* 18 (3): 255–263.

Davies, M. N. O. and Green, P. R. (1990). Optic flow-field variables trigger landing in hawk but not in pigeon. *Naturwissenschaften* 77: 142.

Dayan, P. and Abbott, L. F. (2001) *Theoretical Neuroscience*. Cambridge, MA: MIT Press.

Dembo, A. (1989). On the capacity of associative memories with linear threshold functions. *IEEE Trans on Info Theo* 35(4): 709–720.

Dennis, M. J. and Yip, J. W. (1978). Formation and elimination of foreign synapses on adult salamander muscle. *J Physiol* 174: 299–310. https://doi.org/10.1113.

Di Chiara, G., Morelli, M., Acquas, E. and Carboni, E. (1992). Functions of dopamine in the extrapyramidal and limbic systems. Clues for the mechanism of drug actions. *Arzneimittelforschung* 42(2A): 231–237.

Ditlevsen, S. and Locherbach, E. (2017). Multi-class oscillating systems of interacting neurons. *Stoch Proc and their Appl* 127: 1840–1869.

Djamshidian, A., O'Sullivan, S.S., Wittmann, B.C., Lees, A.J. and Averbeck, B.B. (2011). Novelty seeking behaviour in Parkinson's disease. *Neuropsychologia* 49(9): 2483–2488.

Dodla, R., Svirskis, G. and Rinzel, J. (2006). Well-timed, brief inhibition can promote spiking: Postinhibitory facilitation. *J. Neurophysiol.* 95(4): 2664–2677.

Doll, C. A. and Broadie, K. (2014). Impaired activity-dependent neural circuit assembly and refinement in autism spectrum disorder genetic models. *Front Cell Neurosci* 8: 30.

Drachman, D. A. (2005). Do we have brain to spare?. *Neurology* 64(12): 2004–2005.

du Sautoy, M. (2003). *The Music of the Primes: Searching to Solve the Greatest Mystery in Mathematics*. New York: HarperCollins.

du Sautoy, M. (2011). *The Number Mysteries*. New York: HarperCollins.

Dudai, Y. (1989). *Neurobiology of Memory*. New York: Oxford University Press.

Ebstein, R.P., Novick, O., Umansky, R., Priel, B., Osher, Y., Blaine, D., Bennett, E.R., Nemanov, L., Katz, M. and Belmaker, R.H. (1996). Dopamine D4 receptor (D4DR) exon III polymorphism associated with the human personality trait of Novelty Seeking. *Nature Genet* 12: 78–80.

Eckenhoff, M. F. and Rakic, P. (1991). A quantative analysis of synaptogenesis in the molecular layer of the dentate gyrus in the resus monkey. *Develop Brain Res* 64: 129–135.

Elson, R. N., Selverstone A. I., Abarbanel, H. D. I. and Rabinovich, M. I. (2002). Inhibitory synchronization of bursting in biological neurons: Dependence on synaptic time constant. *J Neurophysiol* 88: 1166-1176.

Elyadi, S. N. (1999). *Discrete Chaos*. Boca Raton: Chapman & Hall/CRC.

Epsztein, J., Brecht, M. and Lee A. K. (2011). Intracellular determinants of hippocampal CA1 place and silent cell activity in a novel environment. *Neuron* 70: 109–120.

Erdős, P. and Rényi, A. (1959). On random graphs *I* . *Publicat. Math.* 6: 290–297.

Erdős, P. and Rényi, A. (1960). On the evolution of random graphs. *Publ. Math. Inst. Hungar. Acad. Sci.* 5: 17–61.

Espay, A. J. (2010). Management of motor complications in Parkinson disease: Current and emerging therapies. *Neurol Clin* 28: 913–925.

Espay, A.J., Gaines, L. Gupta, R. (2013). Sensory feedback in Parkinson's disease patients with "on"-predominant freezing of gait. *Front Neur* 4: 14.

Espay, A. J., Baram, Y., Dwivedi, A. K., Shukla, R., Gartner, M., Gaines, L., Duker, A. P. and Revilla, F. J. (2010). At-home training with closed-loop augmented-reality cueing device for improvement of gait in patients with Parkinson's disease. *J Rehab Res & Devel (JRRD)* 47 (6): 573–582.

Ezhov, A., Nifanova, A. and Ventura, D. (2000). Quantum associative memory with distributed queries. *Inf Sci Inf Comput Sci* 128(3-4): 271–293.

Fagiolini, M. and Hensch, T. K. (2000). Inhibitory threshold for critical-period activation in primary visual cortex. *Nature* 404: 183–186.

Fairhall, A. L., Lewen, G. D., Bialek, W. and van Steveninck, R. R. R. (2001). Efficiency and ambiguity in an adaptive neural code. *Nature* 412: 787–792.

Faisal, A. A., Selen, L. P. J. and Wolpert, D. M. (2008). Noise in the nervous system. *Nat Rev Neurosci.* 9(4): 292–303.

Fell, J., Roschke, J. and Beckmann, P. (1993). Deterministic chaos and the first positive Lyapunov exponent: A nonlinear analysis of the human electroencephalogram during sleep. *Biol Cybern* 69: 139–146.

Feller, M. B. and Scanziani, M. A. (2005). A precritical period for plasticity in visual cortex. *Curr Opin Neurobiol* 15(1): 94–100.

Field, M. and Golubitsky, M. (2009) *Symmetry in Chaos* (2nd ed.), SIAM.

Fine, B. and Rosenberger, G. (1997). *The Fundamental Theorem of Algebra, Undergraduate Texts in Mathematics.* Berlin: Springer-Verlag.

Furstenberg, H. (1955). On the infinitude of primes. *Amer Math Month* 62(5): 353.

Gage, F. H. (2002). Neurogenesis in the adult brain. *J Neurosci* 22(3): 612–613.

Garner, A. and Mayford, M. (2012). New approaches to neural circuits in behavior. *Learn & Mem.* 19: 385–390.

Gerstein, G. L. and Mandelbrot, B. (1964). Random walk models for the spike activity of a single neuron. *Biophys J* 4: 41–68.

Gerstner, W. (1995). Time structure of the activity in neural network models. *Phys Rev E* 51: 738–758.

Gerstner, W. and Kistler, W. M. (2002). *Spiking Neuron Models.* Cambridge University Press.

Gibson, D. A. and Ma, L. (2011). Developmental regulation of axon branching in the vertebrate nervous system. *Development* 138, 183–195.

Gicquel, N., Anderson, J. S. and Kevrekidis, I. G. (1998). Noninvertibility and resonance in discrete-time neural networks for time-series processing. *Phys. Lett. A* 238: 8–18.

Giedd, J. N. *et al.* (1999). Brain development during childhood and adolescence: A longitudinal MRI study. *Nat Neurosci* 2: 861–863.

Gilbert, E. N. (1959). Random graphs. *Annals of Mathematical Statistics* 30(4): 1141–1144.

Goles, E. and Martínez, S. (1990). *Neural and Automata Networks: Dynamical Behavior and Applications.* Kluwer Academic Publishers.

Goold, C. P. and Nicoll, R. A. (2010). Single-cell optogenetic excitation drives homeostatic synaptic depression. *Neuron* 68(3): 512–528.

Graef, S., Biele, G., Krugel, L.K., Marzinzik, F., Wahl, M., Wotka, J., Klostermann, F. and Heekeren, H.R. (2010). Differential influence of levodopa on reward-based learning in Parkinson's disease. *Front Hum Neurosci* 4: 169.

Granit, R., Kernell, D. and Shortess, G. K. (1963). Quantitative aspects of repetitive firing of mammalian motoneurons caused by injected currents. *J Physiol* 168: 911–931.

Griffin, H. J., Greenlaw, R., Limousin, P., Bhatia, K., Quinn, N. P. and Jahanshahi, M. (2011). The effect of real and virtual visual cues on walking in Parkinson's disease. *J Neurol* 258(6): 991–1000.

Grossberg, S. and Todorovic, D. (1988). Neural dynamics of 1-D and 2-D brightness perception: A unified model of classical and recent phenomena. *Percept Psychophys* 43(3): 241–277.

Grover, L. K. (1996). A fast quantum mechanical algorithm for database search. In *STOC '96, Proceedings of the Twenty-eighth Annual ACM Symposium on Theory of Computing*, pp. 212–219.

Groves, P. M., Wilson, C. J., Young, S. J. and Rebe'c, G. V. (1975). Self inhibition by dopaminergic neurons: An alternative to the "neuronal feedback loop" hypothesis for the mode of action of certain psychotropic drugs. *Science* 190: 522–528.

Hafting, T., Fyhn, M., Molden, S., Moser, M. B. and Moser, E. I. (2005). Microstructure of a spatial map in the entorhinal cortex. *Nature* 436(7052): 801–806.

Han, Y., Kebschull, J. M., Campbell, R. A. A., Cowan, D., Imhof, F., Zador, A. M. and Mrsic-Flogel, T. D. (2018). The logic of single-cell projections from visual cortex. *Nature* 556: 51–56.

Hanakawa, T., Fukuyama, H., Katsumi, Y., Honda, M. and Shibasaki, H. (1999). Enhanced lateral pre-motor activity during paradoxical gait in Parkinson's disease. *Ann Neurol* 45(3): 329–336.

Hayashi, H. and Ishizuka, S. (1992). Chaotic nature of bursting discharges in the onchidium pacemaker neuron. *J Theor Biol* 156: 269–291.

Hebb, D. O. (1949). *The Organization of Behavior: A Neuropsychological Theory*. New York: Wiley.

Hensch, T. K. (2005). Critical period plasticity in local cortical circuits. *Nature Reviews Neuroscience* 6: 877–888.

Hensch, T. K., Fagiolini, M., Mataga, N., Stryker, M. P., Baekkeskov, S. and Kash, S. F. (1998). Local GABA circuit control of experience-dependent plasticity in developing visual cortex. *Science* 282(5393): 1504–1508.

Hertz, J., Krogh, A. and Palmer, R. G. (1991). *Introduction to the Theory of Neural Computation*. Reading, MA: Addison Wesley.

Hinton, G. E., McLleland, J. L. and Rumelhart, D. E. (1986). Distributed representations. In *Parallel Distributed Processing (Vol. 1)*. Cambridge, Massuchusetts: MIT Press.

Hodgkin, A. and Huxley, A. A. (1952). Quantitative description of membrane current and its application to conduction and excitation in nerve. *J Physiol* 117: 500–544.

Hooks, B. M. and Chen, C. (2007). Critical periods in the visual system: Changing views for a model of experience-dependent plasticity. *Neuron* 56(2): 312–326.

Hopfield, J. J. (1982). Neural networks and physical systems with emergent collective computational abilities. *Proc Nat Acad Sci USA* 79: 2554–2558.

Horn, D., Levy, N. and Ruppin, E. (1996). Neuronal-based synaptic compensation: A computational study in Alzheimer's disease. *Neur Comp* 8: 1227–1243.

Howe, T. E., Lovgreen, B., Cody, F. W. J., Ashton, V. J. and Oldham, J. A. (2003). Auditory cues can modify the gait of persons with early-stage Parkinson's disease: A method for enhancing Parkinsonian walking performance?. *Clin Rehab* 17: 363–367.

Howell, J. C., Yeazell, J. A. and Ventura, D. (2000). Optically simulating a quantum associative memory. *Physical Review A* 62(4): 042303.

Hsia, A. Y., Malenka, R. C. and Nicoll, R. A. (1998). Development of excitatory circuitry in the Hippocampus. *J Neurophysiol* 79: 2013–2024.

Hu, Y., Trousdale, J., Josić, K. and Shea-Brown, E. (2013). Motif statistics and spike correlations in neuronal networks. *J Stat Mech: Theo & Exper*: P03012; *BMC Neuroscience* (2012), 13(Suppl 1): P43.

Huang, Z. J., Kirkwood, A., Pizzorusso, T., Porciatti, V., Morales, B., Bear, M. F., Maffei, L. and Tonegawa, S. (1999). BDNF regulates the maturation of inhibition and the critical period of plasticity in mouse visual cortex. *Cell* 98(6): 739–755.

Hunt, R. W. (2004). *The Reproduction of Colour* (6th ed). Chichester, UK: Wiley.

Huttenlocher, P. R. (1979). Synaptic density in human frontal cortex. Development changes and effects of age. *Brain Res* 163: 195–205.

Huttenlocher, P. R. and De Courten, C. (1987). The development of synapses in striate cortex of man. *J. Neurosci* 6: 1–9.

Huttenlocher, P. R., De Courten, C., Garey, L. J. and Van der Loos, H. (1982). Synaptogenesis in human visual cortex — evidence for synapse elimination during normal development. *Neurosci Lett* 33: 247–252.

Hyland, B. I., Reynolds, J. N. J., Hay, J., Perk, C. G. and Miller, R. (2002). Firing modes of midbrain dopamine cells in the freely moving rat. *Neuroscie*, 114: 475–492.

Iglesias, J., Eriksson, J., Grize, F., Tomassini, M. and Villa, A. (2005). Dynamics of pruning in simulated large-scale spiking neural networks. *BioSys* 79(9): 11–20.

Ikemoto, S. (2007). Dopamine reward circuitry: two projection systems from the ventral midbrain to the nucleus accumbens-olfactory tubercle complex. *Brain Res Rev* 56(1): 27–78.

Innocenti, G. M. (1995). Exuberant development of connections and its possible permissive role in cortical evolution. *Trends Neurosci* 18: 397–402.

Intrator, N. and Cooper, L. N. (1992). Objective function formulation of the BCM theory of visual cortical plasticity: Statistical connections, stability conditions. *Neur Netw* 5: 3–17.

Itoh, M. and Chua, L.O. (1997). Multiplexing techniques via chaos. In *IEEE Intern. Sym. Circ. Syst.* 2: 905–908.

Ivancevic, V.G. and Ivancevic, T. T. (2010). *Quantum Neural Computation*. New York: Springer.

Izhikevich, E. M. (2000). Neural excitability, spiking and bursting. *International Journal of Bifurcation and Chaos* 10(6): 1171–1266.

Izhikevich, E. M. (2001). Resonate-and-fire neurons. *Neural Networks* 14(6): 883–894.

Jenkinson N. and Brown P. (2011). New insights into the relationship between dopamine, beta oscillations and motor function. *Trends Neurosci* 34(12): 611–618.

Jolivet, R., Lewis, T. J. and Gerstner, W. (2004). Generalized integrate-and-fire models of neuronal activity approximate spike trains of a detailed model to a high degree of accuracy. *J Neurophysiol* 92: 959–976.

Jones, B. R. and Thompson, S. H. (2001). Mechanism of postinhibitory rebound in molluscan neurons. *Integrat Compar Biol.* 41(4): 1036–1048.

Kalaska, J. F., Cohen, D. A., Hyde M. L. and Prud'homme M. (1989). A comparison of movement direction-related versus load direction-related activity in primate motor cortex, using a two-dimensional reaching task. *J Neurosci* 9: 2080–2102.

Kanerva, P. (1988). *Sparse Distributed Memory*. Cambridge, Massachusetts: MIT Press.

Karlsson, M. P. and Frank L. M. (2009). Awake replay of remote experiences in the hippocampus. *Nat Neurosci* 12: 913–918.

Katz, L. C. and Shatz, C. J. (1996). Synaptic activity and the construction of cortical circuits. *Science* 274: 1133–1138.

Kayser, C., Montemurro, M. A., Logothetis, N. K. and Panzeri, S. (2009). Spike-phase coding boosts and stabilizes the information carried by spatial and temporal spike patterns. *Neuron* 61: 597–608.

Keller, T., Piazzo, L., Mandarini, P. and Hanzo, L. (2001). Orthogonal frequency division multiplex synchronization techniques for frequency-selective fading channels. *IEEE J. Select. Ar. in Comm.* 19(6): 999–1008.

Kenet, T., Bibitchkov, D., Tsodyks, M., Grinvald, A. and Arieli, A. (2003). Spontaneously emerging cortical representations of visual attributes. *Nature* 425: 954–956.

Kerchner, G. A. and Nicoll, R. A. (2008). Silent synapses and the emergence of a postsynaptic mechanism for LTP. *Nat Rev Neurosci* 9(11): 813.

Kirtland, J. (2001). *Identification Numbers and Check Digit Schemes*, pp. 43–44. Math Assoc of Amer: MAA Press.

Knoblauch, A. and Sommer, F. T. (2016). Structural Plasticity, Effectual Connectivity, and Memory in Cortex. *Front Neuroanat* 10: 63 doi: 10.3389/fnana.2016.00063.

Knoebel, R. A. (1981). Exponential reiterated. *The Amer Math Month* 88: 235–252.

Koenderink, J. J. and van Doorn, A. J. (1986). Depth and shape from differential perspective in the presence of bending deformations. *J Opt Soc of Amer A* 3(2): 242–249.

Koenigs, G. (1884). Rescherches sur les integrals de certains equations fonctionnelles. *Annales Scientifiques de l'Ecole Normale Superieure* (3)1: 3–41.

Komiyama, T., Sato, T. R., O'Connor, D. H., Zhang, Y. X., Huber, D., Hooks, B. M., Gabitto, M. and Svoboda, K. (2010). Learning-related fine-scale specificity imaged in motor cortex circuits of behaving mice. *Nature* 464: 1182–1186.

Kritikos, A., Leahy, C., Bradshaw, J. L., Iansek, R., Phillips, J. G. and Bradshaw, J. A. (1995). Contingent and non-contingent auditory cueing in Parkinson's disease. *Neuropsych* 33(10): 1193–1203.

Křížek, M., Luca, F. and Somer, L. (2001). *Lectures on Fermat Numbers: From Number Theory to Geometry*, pp. 1–2. New York: Springer-Verlag.

Kuh, A. and Dickinson, B. W. (1989). Information capacity of associative memories. *IEEE Trans Info Theo* 35(1): 59–68.

Kuo, B. C. (1981). *Automatic Control Systems.* Englewood Cliffs, New Jersey: Prentice-Hall.

Kuznetsov, A. S., Kopell, N. J. and Wilson, C. J. (2006). Transient high-frequency firing in a coupled-oscillator model of the mesencephalic dopaminergic neuron. *J Neurophys* 95(2): 932–947.

Lambert, J. D. (1992). *Numerical Methods for Ordinary Differential Systems.* New York: Wiley.

Langton, C. G. (1990). Computation at the edge of chaos: Phase transition and emergent computation. *Physica D* 42: 12–37.

Lankheet, M. J. M., Klink, P. C., Borghius, B. G. and Noest, A. J. (2012). Spike-interval triggered averaging reveals a quasi-periodic spiking alternative for stochastic resonance in catfish electroreceptors. *PLoS ONE* 7(3): e32786.

Lapicque, L. (1907). Recherches quantitatives sur l'excitation électrique des nerfs traitée comme une polarisation. *J Physiol Pathol Gen* 9: 620–635.

Laroche, P., Woltman, G., Blosser, A. *et al.* (2018). Mersenne Primes: History, Theorems and Lists. Available at https://primes.utm.edu/mersenne/index.html (accessed 30 Jan 2021).

Lee, Y. I., Li, Y., Mikesh, M., Smith, I., Nave, K-A., Schwab, M. S. and Thompson, W. J. (2016). Neuregulin1 displayed on motor axons regulates terminal Schwann cell-mediated synapse elimination at developing neuromuscular junctions. *PNAS* 113(4): E479–E487.

Lehmer, E. (1936) On the magnitude of the coefficients of the cyclotomic polynomial. *Bullet of the Amer Math Soc* 42(6): 389–392.

Leloup, J.-C., Gonze, D. and Goldbeter, A. (1999). Limit cycle models for circadian rhythms based on transcriptional regulation in Drosophila and Neurospora. *J. Biol Rhyth* 14(6): 433–448.

Lemeray, E. M. (1895). Sur les fonctions iteratives et sur une nouvelle fonctions. *Association Francaise pour l'Advencement des Sciences, Congres Bordeuaux* 2: 149–165.

Lenstra, H. W. (1987). Factoring integers with elliptic curves. *Ann of Math* 126: 649–673.

Lenstra, H. W. and Pomerance, C. (2002). Primality testing with Gaussian periods. In Agrawal M. and Seth A. (eds), *Foundations of Software Technology and Theoretical Computer Science,* pp.1–1. Berlin: Springer.

Lewis, G. N., Byblow, W. D. and Walt, S. E. (2000). Stride length regulation in Parkinson's disease: The use of extrinsic, visual cues. *Brain* 123: 2077–2090.

L'Huillier, S. A. J. (1812–1813). Mémoire sur la polyèdrométrie. *Annales de Mathématiques* 3: 169–189.

Li, L., Gu, H., Yang, M. Y., Liu, Z. and Ren, W. (2004). A series of bifurcation scenarios in the firing pattern transitions in an experimental neural pacemaker. *Int J Bifur Chaos* 14: 1813–1817.

Li, T.-Y. and Yorke, J. A. (1975). Period three implies chaos. *Am Math Month* 82: 985–992.

Li, Y. and Stuber, G., Eds. (2010). *Orthogonal Frequency Division Multiplexing for Wireless Communications.* New York: Springer.

Liao, D., Zhang, X., O'Brien, R., Ehlers, M. D. and Huganir, R. L. (1999). Regulation of morphological postsynaptic silent synapses in developing hippocampal neurons. *Nat Neurosci* 2(1): 37–43.

Lin, J.-S. (2001). Annealed chaotic neural network with nonlinear self-feedback and its application to clustering problem, *Pattern Recog.* 34: 1093–1104.

Lin, W. L., Ruen, J. and Zhao, W. (2002). On the mathematical clarification of the snap-back repeller in high dimensional systems and chaos in a discrete neural network model. *Int. J. Bifur. & Chaos* 12(5): 1129–1139.

Lisman, J. E. and Grace, A. A. (2005). The hippocampal-VTA loop: controlling the entry of information into long-term memory. *Neuron* 46(5):703–713.

Liu, W., Shi, H., Wang, L. and Zurada, J. M. (2006). Chaotic cellular neural networks with negative self-feedback. In *Artific. Intell. Soft Comp. – ICAISC 2006, Lecture Notes in Computer Science* 4029/2006: 66–75.

Lopes da Silva, F. (1991). Neural mechanisms underlying brain waves: From neural membranes to networks. *Electroencephal and Clin Neurophysiol* 79(2): 81–93.

Losi, G., Prybylowski, K., Fu, Z., Luo, J. H. and Vicini, S. (2002). Silent synapses in developing cerebellar granule neurons. *J Neurophysiol* 87(3): 1263–1270.

Lundstrom, B. N. and Fairhall, A. L. (2006). Decoding stimulus variance from a distributional neural code of interspike intervals. *J Neurosci* 26: 9030–9037.

Maffei, A. and Turrigiano, G. (2008). The age of plasticity: Developmental regulation of synaptic plasticity in neocortical microcircuits. *Prog Brain Res* 169: 211–223.

Mandl, F. (1988). *Statistical Physics* (2nd ed.). Wiley & Sons.

Marder, E. and Bucher, D. (2001). Central pattern generators and the control of rhythmic movements. *Curr Biol* 11(23): 986–996.

Marder, E., Abbott, L. F., Turrigiano, G. G., Liu, Z. and Golowasch, J. (1996). Memory from the dynamics of intrinsic membrane currents. *Proc Nat Acad Sci USA* 93: 13481–13486.

Markram, H., Toledo-Rodriguez, M., Wang, Y., Gupta, A., Silberberg, G. and Wu, C. (2004). Interneurons of the neocortical inhibitory system. *Nat Rev Neurosci* 5(10): 793–807.

Marr, D. and Hildreth, E. (1980). Theory of edge detection. *Proc R Soc Lond B Biol Sci* 207(1167): 187–217.

Martin, J. P. (1967). Locomotion and the basal ganglia. In Martin JP, (ed). *The Basal Ganglia and Posture*, pp. 20–35. London: Pitman Medical.

Matsumoto, M. and Nishimura, T. (1998). Mersenne Twister: A 623-dimensionally equidistributed uniform pseudo-random number generator. *ACM Trans on Model and Comp Sim* 8(1): 3–30.

May, R. M. (1976). Simple mathematical models with very complicated dynamics. *Nature* 261(5560): 459–467.

McCulloch, W. S. and Pitts, W. (1943). A logical calculus of the idea immanent in nervous activity. *Bullet Math Biophys* 5: 115–133.

McEliece, R. J., Posner, E. C., Rodemich, E. R. and Venkatesh, S. (1987). The capacity of the hopfield associative memory. *IEEE Trans Info Theo* IT-33(4): 461–482.

McGahon, B. M., Martin, D. S., Horrobin, D. F. and Lynch, M. A. (1999). Age-related changes in synaptic function: analysis of the effect of dietary supplementation with omega-3 fatty acids. *Neuroscience* 94(1): 305–14.

McIntosh, G. C., Brown, S. H., Rice, R. R. and Thaut, M. H. (1997). Rhythmic auditory-motor facilitation of gait patterns in patients with Parkinson's disease. *J Neurol Neurosurg Psych* 62: 22–26.

Mechelli, A., Crinion, J. T., Noppeney, U., O'Doherty, J., Ashburner, J., Frackowiak, R. S. J. and Price, C. J. (2004). Neurolinguistics: Structural plasticity in the bilingual brain. *Nature* 431(7010): 757–757. https://doi.org/10.1038/431757a

Melnick, I. V. (1994 Rus, 2010 Eng). Electrically silent neurons in the substantia gelatinosa of the rat spinal cord. *Fiziol Zh* 56(5): 34–39.

Menza, M. A., Golbe, L. I., Cody, R. A. and Forman, N. E. (1993). Dopamine-related personality traits in Parkinsons disease. *Neurology* 43: 505–508.

Miendlarzewska W. J. and Trost E. A. (2014). How musical training affects cognitive development: rhythm, reward and other modulating variables. *Front Neurosci* 20 January.

Miller, G. A. (1956). The magical number seven, plus or minus two: Some limits on our capacity for processing information. *Psychol Rev* 63: 81–97.

Miller, G. L. (1976). Riemann's hypothesis and tests for primality. *J Comp Syst Sci* 13(3): 300–317.

Miller, K. D. and Fumarola, F. (2012). Mathematical equivalence of two common forms of firing rate models of neural networks. *Neural Computation* 24: 25–31.

Ming-Chia, L., Yasuda, R. and Ehlers, M. D. (2010). Metaplasticity at single glutamatergic synapses. *Neuron* 66(6): 859–870.

Minsky, M. and Papert, S. (1969). *Perceptrons: An Introduction to Computational Geometry*. Cambridge: MIT Press.

Mira, C. (2007). Noninvertible maps. *Scholarpedia* 2(9): 2328.

Mitchell, M., Hraber, P. T. and Crutchfield, J. P. (1993). Revisiting the edge of chaos: Evolving cellular automata to perform computations. *Complex Systems* 7: 89–130.

Miyajima, H., Shigei, N. and Arima, K. (2010). Some quantum search algorithms for arbitrary initial amplitude distribution. In *Sixth International Conference on Natural Computation (ICNC)*, 8: 603–608.

Miyata, S., Komatsu, Y., Yoshimura, Y., Taya, C. and Kitagawa, H. (2012). Persistent cortical plasticity by upregulation of chondroitin 6-sulfation. *Nat Neurosci* 15(3): 414–422.

Mizraji, E. and Lin, J. (2017). The feeling of understanding: an exploration with neural models. *Cogn Neurodyn* 11(2): 135–146.

Mollin, R. A. (2002). A brief history of factoring and primality testing B. C. (before computers). *Math Magaz* 75(1): 18–29.

Mongillo, G., Barak, O. and Tsodyks, M. (2008). Synaptic theory of working memory. *Science* 319: 1543–1546.

Mongillo, G., Rumpel, S. and Loewenstein, Y. (2018). Inhibitory connectivity defines the realm of excitatory plasticity. *Nat Neurosci* 21: 1463–1470.

Moreno, J. L. and Jennings, H. H. (1938). Statistics of social configurations. *Sociometry* 1 (3/4): 342–374.

Morris, M. E., Iansek, R., Matyas, T. A. and Summers, J. J. (1996). Stride length regulation in Parkinson's disease: Normalizations strategies and underlying mechanisms. *Brain* 119: 551–568.

Moustafa, A. (2010). Levodopa enhances reward learning but impairs reversal learning in Parkinson's disease patients. *Front Hum Neurosci* 4: 240.

Murray, J. (2002). *Mathematical Biology: An Introduction, Volume 1* (3rd ed.), p. 551. New York, NY: Springer.

Murthy, V. N. and Fetz, E. E. (1996). Oscillatory activity in sensorimotor cortex of awake monkeys: Synchronization of local field potentials and relation to behavior. *J Neurophysiol* 76: 3949–3967.

Nelson, R. C. and Aloimonos, J. (1989). Obstacle avoidance using flow field divergence. *IEEE Trans on Patt Anal Mach Intell* 11(10): 1102–1106.

Neumann, N. (2010). The mind of the mnemonists: An MEG and neuropsychological study of autistic memory savants. *Behav Brain Res* 215(1): 114–121.

Nielsen, M. A. and Chuang, I. L. (2000). *Quantum Computation and Quantum Information*. Cambridge University Press.

Nieuwboer, A., Kwakkel, G., Rochester, L., Jones, D., van Wegen, E., Willems, A. M., Chavret, F., Hetherington, V., Baker, K. and Lim, I. (2007). Cueing training in the home improves gait-related mobility in Parkinson's disease. The RESCUE trial. *J Neurol Neurosurg & Psych* 78: 134–140.

Nissen, T. and Wynn, R. (2014). The clinical case report: a review of its merits and limitations. *BCM Res Notes* 7: 264.

O'Keefe, J. and Dostrovsky, J. (1971). The hippocampus as a spatial map. Preliminary evidence from unit activity in the freely-moving rat. *Brain Research* 34(1): 171–175.

O'Keefe, J. and Nadel, L. (1978). *The Hippocampus as a Cognitive Map*. Oxford University Press.

Ohta, M. (2002). Chaotic neural networks with reinforced self feedbacks and its application to N-queen problem. *Math. Comp. Sim.*, 59: 305–317.

Oja, E. (1982). A simplified neuron model as a principal component analyzer. *J Math Biol* 15 (3): 267–273.

Ott, E. (1993). *Chaos in Dynamical Systems.* Cambridge University Press.

Panzeri, S., Brunel, N., Logothetis, N. K. and Kayser, C. (2009). Sensory neural codes using multiplexed temporal scales (Review). *Trends in Neurosci* 33: 111–120.

Pattersen, K. H. and Einevoll, G. T. (2008). Amplitude variability and extracellular low-pass filtering of neuronal spikes. *Biophys J* 94: 784–802.

Penrose, R. (1994). *Shadows of the Mind.* Oxford University Press.

Perkel, D. H. and Mulloney, B. (1974). Motor pattern production in reciprocally inhibitory neurons exhibiting postinhibitory rebound. *Science* 185: 181–182.

Peterfreund, N. and Baram, Y. (1994a). Second–order bounds on the domain of attraction and the rate of convergence of nonlinear dynamical systems and neural networks. *IEEE Trans on Neur Net* 5(4): 551–560.

Peterfreund, N. and Baram, Y. (1994b). Trajectory control of convergent networks. *Neur Proc Lett* 8: 99–106.

Peterson, I. (1999). The Return of Zeta. Available at: https://static1.squarespace.com/static/54c161ffe4b063fc8ab03446/t/54cab277e4b042fd7653bec5/1422570103763/The+Return+of+Zeta.pdf (accessed 30 Jan 2021).

Phillips, M. A., Colonnese, M. T., Goldberg, J., Lewis, L. D., Brown, E. N. and Constantine-Paton, M. A. (2011). Synaptic strategy for consolidation of convergent visuotopic maps. *Neuron* 71(4): 710–724.

Pollard, J. M. (1975). A Monte Carlo method for factorization. *BIT Numer. Math.* 15(3): 331–334.

Pomerance, C. (1996). A tale of two sieves. *Notices of the AMS* 43(12): 1473–1485.

Poon, C.-S., Young, D. L. and Siniaia, M. (2000). High-pass filtering of carotid-vagal influences on expiration in rat: Role of N-methyl-D-aspartate receptors. *Neurosci Let* 284: 5–8.

Pratt, V. (1975) Every prime has a succinct certificate. *SIAM Comput* 4: 214–220.

Prazdny, K. (1983). On the information in optical flows. *Comp Vis, Graphics, Image Proc* 22(2): 239–259.

Pribram, K. H., Miller, G. A. and Galanter, E. (1960). Plans and the Structure of Behavior, p. 65. New York: Holt, Rinehart and Winston.

Prothero, J. (1993). The treatment of akinesia using virtual images. M.S. Thesis, Human Interface Technol Lab, Univ Washington, Seattle.

Qwakenaak, H. and Sivan, R. (1972). *Linear Optimal Control Systems.* New York: Wiley Interscience.

Rabin, M. O. (1980). Probabilistic algorithm for testing primality. *J Numb Theo* 12(1): 128–138.

Rakic, P., Bourgeois, J. P. and Goldman-Rakic, P. S. (1994). Synaptic development of the cerebral cortex: Implications for learning, memory and mental illness. *Prog in Brain Res* 102: 227–243.

Rao, A. R. (2018). An oscillatory neural network model that demonstrates the benefits of multisensory learning. *Cogn Neurodyn* 12(5): 481–499.

Ren, W. S., Hu, J., Zhang, B. J., Wang, F. Z., Gong, Y. F. and Xu, J. X. (1997). Period-adding bifurcation with chaos in the interspike intervals generated by an experimental neural pacemaker. *Int J Bifur Chaos* 7: 1867–1872.

Ribenboim, P. (2017). *Prime Numbers, Friends Who Give Problems.* New Jersey: World Scientific.

Riesel, H. (1994). *Prime Numbers and Computer Methods for Factorization.* Basel: Birkhauser.

Riess, T. and Weghorst, S. (1995). Augmented reality in the treatment of Parkinson's disease. In Morgan, K., Satava, R. M., Sieburg, H. B., Mattheus, R. and Christensen, J. P. (eds). *Proc Medicine Meets Virtual Reality III.*, pp. 298–302. Amsterdam, The Netherlands: IOS Press.

Ringach, D. L. and Baram, Y. (1992). Obstacle detection by diffusion. CIS Report 9220, Computer Science Dept., Technion, Israel Institute of Technology.

Ringach, D. L. and Baram, Y. (1994). A diffusion mechanism for obstacle detection from size–change information. *IEEE Trans on Patt Anal Mach Intell* 16(1): 76–80.

Rosen, R. (1958). The representation of biological systems from the standpoint of the theory of categories. *Bullet of Math Biophys* 20: 317–341.

Rosenblatt, F. (1958). The perceptron: A probabilistic model for information storage and organization in the brain. *Psychol. Rev* 65(6): 386–408.

Roth, Z. and Baram, Y. (1996). Multi–dimensional density shaping by sigmoids. *IEEE Trans. on Neur Netw* 7(5): 1291–1298.

Salimpoor, V. N., Zald, D. H., Zatorre, R. J., Dagher, A. and McIntosh, A. R. (2015). Predictions and the brain: how musical sounds become rewarding. *Trends Cogn Sci* 19(2): 86–91.

Salman T. and Baram Y. (2012). Quantum Set Intersection and its Application to Associative Memory. *J Machine Learning Research* 13: 3177–3206.

Sandifer, C. E. (2014). *How Euler Did Even More.* Cambridge University Press.

Sato, Y., Akiyama, E. and Farmer, J. D. (2002). Chaos in learning a simple two-person game. *Proc. of the Nat. Acad. Sci. (PNAS)* 99(7): 4748–4751.

Schelter, B., Timmer, J. and Micahel, E. (2009). Assessing the strength of directed influences among neural signals using renormalized partial directed coherence. *J Neu Meth* 179(1):121–130.

Seborg D. E., Mellichamp, D. A. and Edgar, T. F., F. J. III. (2011). *Process Dynamics and Control* (2nd ed.). New York: Wiley.

Sharp, A. A., Skinner, F. K. and Marder, E. (1996). Mechanism of oscillation in dynamic clamp constructed two-cell half-center circuits. *J. Neurophys* 76: 867–883.

Shilnikov, A. L. and Rulkov, N. F. (2003). Origin of chaos in a two-dimensional map modeling spiking-bursting neural activity. *Int. J. Bifur. & Chaos* 13(11): 3325–3340.

Shor, P. W. (1994). Algorithms for quantum computation: Discrete logarithms and factoring. In *Proceedings of the IEEE Symposium on Foundations of Computer Science*: 124–134.

Smith, T. C. and Jahr, C. E. (2002). Self-inhibition of olfactory bulb neurons. *Nature Neurosci* 5: 760–766.

Snijders, A., Leunissen, I., Bakker, M., Overeem, S., Hemlich, R. C., Bloem, B. R. and Toni, I. (2011). Gait-related cerebral alterations in patients with Parkinson's disease with freezing of gait. *Brain* 134: 59–72.

So, P., Francis, J. T., Netoff, T. I., Gluckman, B. J. and Schiff, J. (1998). Periodic orbits: A new language for neuronal dynamics. *Biophys J* 74: 2776–2785.

Solomon, J. (2006). Case studies: Why are they so important. *Nature Clinical Practice Cardiovascular Medicine* 3: 579.

Solomonoff, R. and Rapoport, A. (1951). Connectivity of random nets. *Bulletin of Mathematical Biophysics* 13(2): 107–117.

Souza, C. dO., Mariana Callil Voos, M. C., Chien, H. F., Bran, R., Barbosa, A. F., Fonoff, F. C., Caromano, F. A., de Abreu, L. C., Barbosa, E. R. and Fonoff, E. T. (2015). Combined auditory and visual cueing provided by eyeglasses influence gait performance in Parkinson's disease patients: A pilot study. *Internat Arch Medic Sect: Neurology* 8(132): 1–8.

Sprott, J. C. (2003). *Chaos and Time-Series Analysis*. Oxford University Press.

Stein, R. B., Gossen, E. R. and Jones, K. E. (2005). Neuronal variability: noise or part of the signal?. *Nat Rev Neurosci* 6(5): 389–397.

Stillwell, J. (2010). *Mathematics and Its History. Undergraduate Texts in Mathematics* (3rd ed.), p. 40. Springer.

Stratton P. and Wiles J. (2015). Global segregation of cortical activity and metastable dynamics. *Front Syst Neurosci* 25(9): 119.

Tessier, C. R. and Broadie, K. (2009). Activity-dependent modulation of neural circuit synaptic connectivity. *Front Mol Neurosci* 30: 2–8.

Thaut, M. H., McIntosh, G. C. and Hoemberg, V. (2014). Neurobiological foundations of neurologic music therapy: rhythmic entrainment and the motor system. *Front Psychol* 5: 1185.

Thaut, M. H., McIntosh, G. C., Rice, R. R., Miller, R. A., Rathbun, J. and Brault, J. M. (1996). Rhythmic auditory stimulation in gait training for Parkinson's disease patients. *Move Disord* 11: 193–200.

The Free Dictionary. (2018). Available at: http://www.thefreedictionary.com (accessed 30 Jan 2021).

Tong, J., Kong, C., Wang, X., Liu, H., Li, B. and He, Y. (2019). Transcranial direct current stimulation influences bilingual language control mechanism: evidence from cross-frequency coupling. *Cogn Neurodyn*. https://doi.org/10.1007/s11571-019-09561-w:1-12

Trudeau, R. J. (1993). *Introduction to Graph Theory*. Dover Books on Mathematics.

Tsodyks, M. and Feigel'man, M. V. (1988). The enhanced storage capacity in neural networks with low activity level. *EPL (Europhysics Letters)* 6(2):101.

Turova, T. S. (2010). The largest component in subcritical inhomogeneous random graphs. *Combin., Prob. Comp.* 1–24.

Van der Loos, H. and Glaser, E. M. (1972). Autapses in neocortex cerebri: Synapses between a pyramidal cell's axon and its own dendrites. *Brain Research* 48: 355–360.

van Praag, H., Christie, H. B., Sejnowski, T. J. and Gage F. H. (1999). Running enhances neurogenesis, learning, and long-term potentiation in mice. *PNAS* 96(23): 13427–13431.

Vanderhaeghen, P. and Cheng, H. J. (2010). Guidance molecules in axon pruning and cell death. *Cold Spring Harbor Perspect in Biol* 2(6): 1–18.

Vardi, R., Wallach, A., Kopelowitz, E., Abeles, M., Marom, S. and Kanter, I. (2012). Synthetic reverberating activity patterns embedded in networks of cortical neurons. *EPL* 97: 66002. doi: 10.1209/0295-5075/97/66002.

Velu, P. D., Mullen, T., Noh, E., Valdivia, M., Poizner, H., Baram, Y. and de Sa, V. R. (2013). Effect of visual feedback on the occipito-parietal-motor network in Parkinson's disease patients with freezing of gait. *Frontiers in Neurology* 4(209).

Ventura, D. and Martinez, T. (2000). Quantum associative memory. *Info Sci* 124(1-4): 273–296.

Verschure, P. F. M. J. (1991). Chaos-based learning. *Complex Systems* 5: 359–370.

Vincente, C. J. P. and Amit, D. A. (1989). Optimised network for sparsely coded patterns. *Journal of Physics A* 22: 559–569.

von Bartheld, C. S., Bahney, J. and Herculano-Houzel, S. (2016). The search for true numbers of neurons and glial cells in the human brain: A review of 150 years of cell counting. *J. Compar. Neurol.* 524: 3865–3895.

Wang, L., Fontanini, A. and Maffei, A. (2012). Experience-dependent switch in sign and mechanisms for plasticity in layer 4 of primary visual cortex. *J Neurosci* 32(31): 10562–10573.

Wang, X. J. (2010). Neurophysiological and computational principles of cortical rhythms in cognition. *Physiol Rev* 90(3): 1195–1268.

Wark, B., Fairhall, A. and Rieke, F. (2009). Timescales of inference in visual adaptation. *Neuron* 61: 750–761.

Wassum, K. M., Ostlund, S. B., Balleine, B. W. and Maidment, N. T. (2011). Differential dependence of Pavlovian incentive motivation and instrumental incentive learning processes on dopamine signaling. *Learn Mem* 18(7): 475–483.

Weghorst, S., Prothero, J. and Furness, T. (1994). Virtual images in the treatment of Parkinson's disease akinesia. In *Proc Medicine Meets Virtual Reality II*, pp. 242–243.

Wei H., Dai D. and Bu Y. (2017). A plausible neural circuit for decision making and its formation based on reinforcement learning. *Cogn Neurodyn* 11(3): 259–281.

Weiner, J. A., Burgess, R. W. and Jontes J, eds. (2013). Mechanisms of neural circuit formation. *Front Mol Neurosci* 6: 12.

Williams, K. S. and C. Simon (1995). The ecology, behavior, and evolution of periodical cicadas. *Ann Rev Entomolog* 40: 269–295.

Wilson, H. R. and Cowan, J. D. (1972). Excitatory and inhibitory interactions in localized populations of model neurons. *Biophys J*12: 1–24.

Wright, J. (1984). Method for calculating a Lyapunov exponent. *Phys. Rev. A* 29: 2924.

Yaro, C. and Ward, J. (2007). Searching for Shereshevskii: What is superior about the memory of synaesthetes?. *Quart J Exper Psych* 60(5): 681–695.

Zlochin M. and Baram, Y. (2001). Manifold stochastic dynamics for Bayesian learning. *Neur Comput* 13(11): 2549–2572.

Index

visual information, 216
voluntary action, 234

walking over transverse lines, 227
walking pace, 220

wearable sensory feedback device, 230
words, 37, 88
working memory, 68, 85, 88, 174
working memory capacity, 54